Oil & Gas Production

Oil & Gas Production

Editor

Sanjay Jadhav

scitus
academics

Oil & Gas Production

Edited by **Sanjay Jadhav**

Printed in 2017

ISBN: 978-1-68117-425-9

Library of Congress Control Number: 2015936540

© 2016 by

SCITUS Academics LLC,
616, Corporate Way, Suite 2, 4766,
Valley Cottage, NY 10989

www.scitusacademics.com

Contents

Preface

This book has been compiled for readers with an interest in the oil and gas industry. It is an overview of the main processes and equipment. When we searched for a suitable introduction to be used for new engineers, I discovered that much of the equipment is described in standards, equipment manuals and project documentation. Little material was found to quickly give the reader an overview of the entire oil and gas industry, while still preserving enough detail to let the engineer have an appreciation of the main characteristics and design issues. I have had many requests that downstream processes be included, and have restructured the book into Upstream, Midstream, Refining and Petrochemical, adding basic information on these facilities.

Editor

Chapter 1

Bio Gas Oil Production from Waste Lard

Jenő Hancsók, Péter Baladincz, Tamás Kasza, Sándor Kovács, Csaba Tóth, andZoltán Varga

MOL Department of Hydrocarbon and Coal Processing, University of Pannonia, H-8201 Veszprém, Hungary

ABSTRACT

Besides the second generations bio fuels, one of the most promising products is the bio gas oil, which is a high iso-paraffin containing fuel, which could be produced by the catalytic hydrogenation of different triglycerides. To broaden the feedstock of the bio gas oil the catalytic hydrogenation of waste lard over sulphided NiMo/Al$_2$O$_3$ catalyst, and as the second step, the isomerization of the produced normal paraffin rich mixture (intermediate product) over Pt/SAPO-11 catalyst was investigated. It was found that both the hydrogenation and the decarboxylation/decarbonylation oxygen removing reactions took place but their ratio depended on the process parameters (T = 280–380°C, P = 20–80 bar, LHSV = 0.75–3.0 h^{-1} and H$_2$/lard ratio: 600 Nm^3/m^3). In case of the isomerization at the favourable process parameters (T = 360–370°C, P = 40 –50 bar, LHSV = 1.0 h^{-1} and H$_2$/hydrocarbon ratio: 400 Nm^3/m^3) mainly mono-branching isoparaffins were obtained. The obtained products are excellent Diesel fuel blending components, which are practically free of heteroatoms.

INTRODUCTION

Nowadays the production and utilisation of agricultural origin fuels have come to the front because of environmental protection, political and economical reasons. In the increase of the utilisation of these bio-fuels the energy policy of the European Union has a dominant role, of which the aim is the decrease of the energy and the crude oil dependency. In consequence, the EU implemented the 2003/30/EC and the 2009/28/EC directives to inspire the utilisation of bio-fuels by determining the suggested and specified ratio of bio-fuels in the transportation fuels. In order to attain these objectives, the fuel blending components produced from different natural triglyceride containing feedstocks (like vegetable oils, used frying oils, animal fats, algae oils, trap grease of sewage works) could have an important role.

Currently only the biodiesels (fatty acid methyl esters) belonging to the first generation bio-fuels are used as bio-origin fuel or a blending component for Diesel engines. But the production and utilisation of these have numerous disadvantages (formation of hazardous wastes, difficulties of glycerol sales, high content of olefinic double bond → poor heat and oxidation stability → poor storage stability; ester bonds and high water content → hydrolysis sensitivity → corrosion; phosphor content → poisonous effect on the three-way catalyst, etc.) [1–3].

Because of these disadvantages and the demand for better quality, the production possibilities of biofuels for diesel engines which have different chemical structure, consequently dissimilar service properties are keenly researched. The research, production and utilization of this kind of second generation biofuels are supported with high priority by the European Union. Besides these the so-called bio gas oils could come to the front in the short and middle-term. The bio gas oils are the mixtures of gas oil boiling point range hydrocarbons (mainly >99% normal and isoparaffins) produced from mainly triglyceride containing feedstocks (vegetable oils, used frying oils, animal fats, etc.) by heterogenic catalytic hydrogenation in one or more stages. The bio gas oils eliminate every disadvantage of the biodiesels (fatty acid esters), accordingly their economical production with high yield could have high importance in the future [4, 5].

The mixtures of the gas oil boiling point range hydrocarbons produced by the catalytic hydrogenation of triglycerides contain mainly normal paraffins. They have outstanding cetane number (>85), but their cold flow properties (e.g., cold filter plugging point) are unfavourable (e.g., the freezing points of the C_{16}–C_{18} normal paraffins are between +18°C and +28°C). For this reason, the chemical structure of the n-paraffins has to be modified to be excellent gas oil blending components. For this purpose, the most suitable technical solution is the catalytic hydroisomerisation. During the isomerization the normal paraffins having high freezing point and outstanding cetane number could be converted to isoparaffins having by far lower freezing points and still high cetane number [4, 5].

Accordingly, the mentioned two-stage catalytic conversion of the triglycerides of different vegetable oils to bio gas oil is discussed in several publications. But there is only a little information about the utilisation of waste fats. Because of these, and to broaden the range of the feedstock of the bio gas oils, we investigated the two-stage catalytic transformation of lard produced from slaughterhouse waste.

The aim of our experimental work was the investigation of the fuel purpose convertibility (to good quality gasoil boiling range product with high isomer content) of properly pretreated Hungarian lard produced from slaughter house waste over an expediently chosen $NiMo/Al_2O_3$ then on a Pt/SAPO-11 catalyst. The experiments were carried out in a laboratory scale reactor system in continuous operation, while the activity of the catalyst, the yield and composition of the products, the possible reaction routes of the deoxygenation and the quantity of isomers, as well as the applicability of the products were investigated in function of the process parameters (temperature, pressure, liquid hourly space velocity, H_2/hydrocarbon volume ratio).

EXPERIMENTAL PART

Feed stocks

The feedstock of the hydrogenation experiments was properly pretreated (filtered, purified by bleaching earth) lard which was produced from slaughter house wastes. Its important properties are summarized in Table 1. For the catalytic hydrogenation of the feedstock, the applied catalyst was a conventional hydrotreating NiMo/Al_2O_3 catalyst. The Ni-content of the catalyst was 3.23% and the Mo-content was 13.4%; the BET surface area was 214 m²/g and the acidity was 0.489 mmol NH_3/g. The catalyst was presulphided before the experiments with previously deep desulphurized gas oil having enhanced (2.5% with dimethyl-disulphide) sulphur content. To maintain the sulphided form of the catalyst, the sulphur content of the feedstock was adjusted to 1000 mg/kg with dimethyl-disulphide. This compound easily dissociates in the applied temperature range. For the isomerization of the intermediate paraffin rich mixture the catalyst was a 0.5% Pt/SAPO-11 catalyst. The dispersion of the platinum was 91%, the BET surface area was 100.1 m²/g, and the acidity of the catalyst was 0.66 mmol NH_3/g. prior to the activity measurements, the catalysts were pre-treated in situ, as described in our earlier publication [5].

Table 1: The main properties and fatty acid composition of the applied waste lard

Properties	Value	Standard methods
Kinematic viscosity, 40°C, mm2/s	39.53	EN ISO 3104:1996
Density (40°C), g/cm3	0.9385	EN ISO 3675: 2000
Sulphur content, mg/kg	6	EN ISO 20846:2004
Acid number, mg KOH/g	0.63	EN 14104:2004
Iodine number, g I2/100 g	78	EN 14111:2004
Carbon residue, %	0.11	EN ISO 10370:1997

Flash point, °C	>250	EN ISO 2719:2003
Cold filter plugging point, °C	37	EN 116:1999
Fatty acid composition*, %		
C14:0	1.04	EN ISO 5509:2000;
C16:0	20.47	EN 14103:2004
C16:1	7.65	
C18:0	12.83	
C18:1	32.99	
C18:2	17.32	
C18:3	6.70	
C20:x	0.86	
Other	0.14	
Calculated oxygen content, %	11.2	

*The first number represents the number of carbon atoms and the second means the number of double bonds in the molecule.

Experimental Apparatus and Product Separation

The experimental tests were carried out in a high pressure reactor system containing two tubular reactors with effective catalyst volume of $100\,cm^3$ [5]. The reactor system contained all the equipment and devices applied in the reactor system of a hydrogenation and isomerization plant. The apparatus was suitable to keep the major process parameters with at least such precision as used in the industry.

The intermediate product mixtures obtained from the hydrogenation of the waste lard were separated to gas phase, water phase and liquid organic phase (hereafter organic phase) (Figure 1). The gas phase obtained from the separator of the reactor system contained mainly hydrogen, carbon-monoxide, carbon-dioxide, propane, hydrogen-sulphide, ammonia which evolved during the heteroatom removal of the triglyceride molecules, furthermore the lighter hydrocarbons (C_1–C_4 as valuable by-products) which formed during the hydrocracking reactions. The liquid product mixtures obtained from the separator of the reactor system contained water, hydrocarbons and oxygen containing compounds. After the separation of water, we obtained the light (C_5–C_9) hydrocarbons (gasoline boiling range) from the organic fraction by distillation up to 180°C.

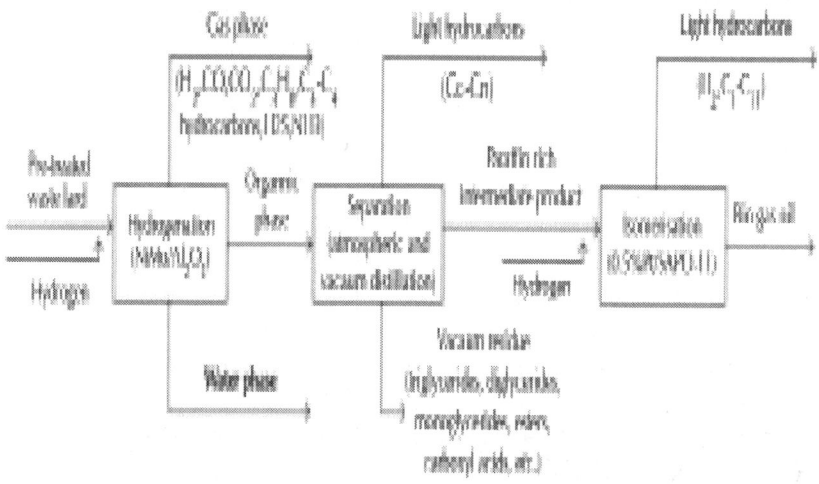

Figure 1: The scheme of the laboratory scale production of the bio gas oil.

The residue of atmospheric distillation was separated by vacuum distillation into the intermediate product (gas oil boiling range fraction, mainly C_{11}–C_{19} hydrocarbons) and the residue. The residue contained the unconverted triglycerides, the evolved and unconverted diglycerides and monoglycerides, furthermore fatty acids and esters, which evolved as intermediate products or were originally in the feedstock. After the isomerization of the normal paraffin rich intermediate product the bio gas oil and some light hydrocarbons were obtained.

Analytical Methods and Calculation Methods

The properties of the lard feedstock, the hydrogenated intermediate products and the target products were measured according to standard methods (see Table 1).

The composition of the organic product mixtures, obtained from the catalytic conversion of triglycerides was determined by high temperature gas chromatograph (Shimadzu 2010 GC) at the following measurement parameters:

- Zebron DB-1HT (30 m × 0.32 mm × 0.1 μm) column,
- PTV (Programmed Temperature Vaporization) injector (temperature program: 100°C → 400°C, 30°C/min heating rate, then 18 min at 400°C),
- oven temperature program: 40°C (4 min) → 240°C, 15°C/min heating rate → 400°C 8°C/min heating rate, then 11 min at 400°C,
- FID (flame ionization) detector (400°C),
- Carrier gas: H_2 (5.0), pressure 58.0 kPa, column flow 4.00 cm³/min.

The nickel and molybdenum content of the hydrogenation catalyst was measured with ICP-OES method. The SAPO-11 catalysts were prepared as described and characterized according to HU 225 912 patent [6]. The platinum content of the isomerization catalyst was determined according to UOP-274 standard. The dispersity of the platinum was determined by H_2 chemisorption [7]. The surface properties of the catalyst were investigated by ASAP 2000 (Micromeritics) equipment (pore diameter in the range of 1.7–300 nm) and by mercury penetration method with CARLO-ERBA equipment (pore diameter in the range of 7.5–15000 nm). The surface area of the catalysts was calculated by using BET-plots. The acidity of the catalysts was determined with temperature programmed desorption of ammonia (TPD-NH$_3$).

We calculated the conversion of triglycerides by the following equation:

$$\text{Conversion} = \frac{X_{\text{feedstock}} - X_{\text{products}}}{X_{\text{feedstock}}} * 100, \tag{1}$$

Where: $X_{\text{feedstock}}$ triglyceride content of the feedstock and X_{products}: triglyceride content of the products.

RESULTS AND DISCUSSION

The Catalytic Hydrogenation of the Waste Lard

The range of the applied process parameters—based on our pre-experimental results—was the following: temperature 280–380°C, total pressure 20–80 bar, liquid hourly space velocity (LHSV) = 0.75–3.0 h^{-1} and H$_2$/waste lard volume ratio = 600 Nm^3/m^3.

Conversions and Product Yields

The conversion of triglycerides increased significantly by increasing the severity of the process parameters, namely by increasing the temperature and the pressure furthermore by decreasing the LHSV (Figures 2 and3). At every investigated value of pressure, at least 1.5 h^{-1} and lower LHSV and at least 320°C or higher temperature was necessary to reach at least 95% conversion of the triglycerides at the 600 Nm^3/m^3 H$_2$/lard volume ratio, which was found to be favorable according to our pre-experiments.

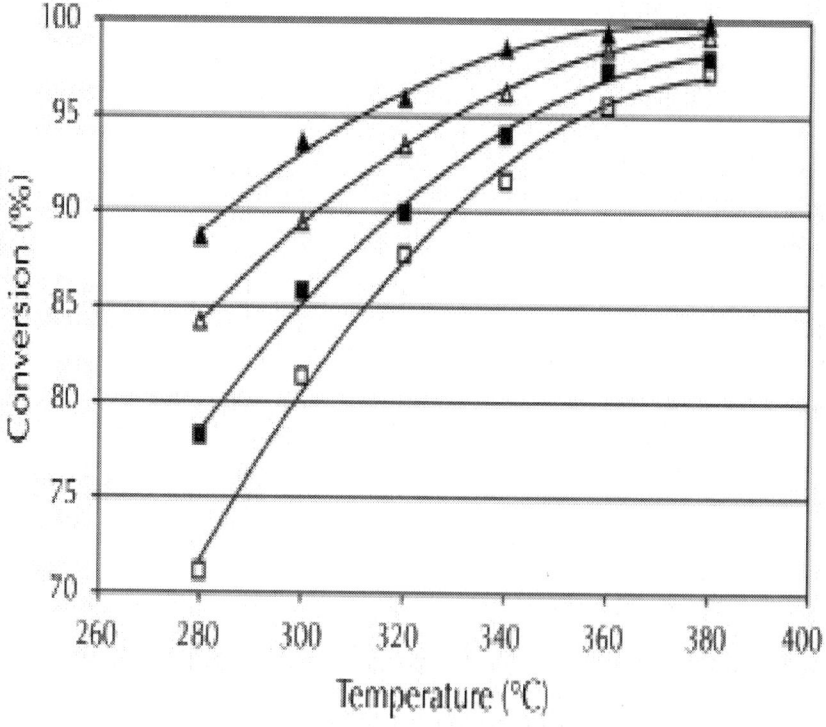

Figure 2: Conversion of triglycerides as a function of the temperature and the pressure; where: □-20 bar, ■-40 bar, -60 bar and ▲-80 bar pressure (LHSV = 1.0 h^{-1}, H$_2$/waste lard volume ratio = 600 Nm^3/m^3).

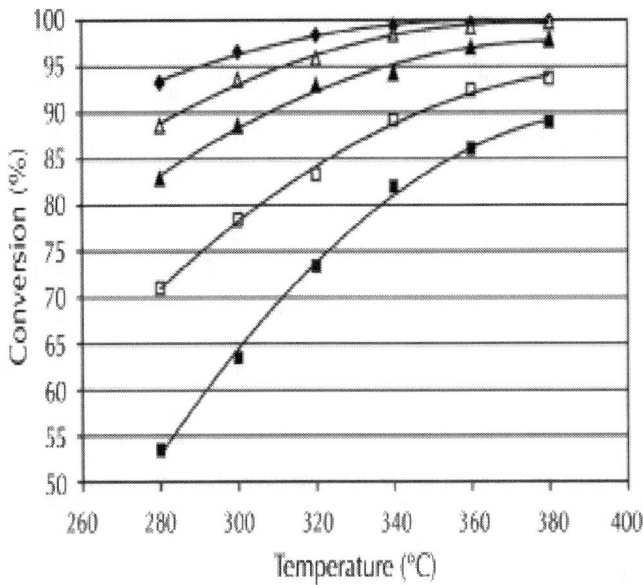

Figure 3: Conversion of triglycerides as a function of the temperature and the LHSV; where: ◆-0.75 h⁻¹, -1.0 h⁻¹, ▲-1.5 h⁻¹, □-2.0 h⁻¹ and ■-3.0 h⁻¹ of LHSV (P = 80 bar, H_2/waste lard volume ratio = 600 Nm³/m³).

The yield of the intermediate normal paraffin-rich product (INPRP) in the gas oil boiling range was significantly influenced by the temperature, the pressure and the LHSV (Figures 4 and 5). At every value of pressure and LHSV, up to 320°C, the yield of the intermediate product increased by increasing the temperature in accordance with the increasing rate of triglyceride conversion. But when the rate of conversion exceeded 95%–98%, then the further increase of temperature caused lower amount of plus paraffins than the amount of hydrocarbon which was decomposed by cracking; consequently, the yield started to decrease. So the yield of INPRP changed according to a maximum curve as a function of the temperature at LHSV of 0.75 and 1.0 h⁻¹, because in these cases the contact time of the hydrocarbons and oxygenic compounds generated from the triglycerides was too long; consequently, by increasing the temperature above 320°C (LHSV = 0.75 h⁻¹) and 360°C (LHSV = 1.0 h⁻¹), the yield decreased because of the increasing rate of cracking. The low yield of INPRP at LHSV of 1.5–3.0 h⁻¹ was caused unequivocally by the low conversion of the triglycerides, as these molecules could not contact long enough with the active sites of the catalyst. In the investigated parameter range at constant temperature, the higher the pressure was the higher yield of INPRP was reached in accordance with the increasing rate of triglyceride conversion (in spite of the fact that the higher the pressure was, the higher the rate of hydrocracking reactions was, as well). (It should be highlighted that in case of the investigated waste lard having the mentioned composition (Table 1), the maximum yield of the gas oil boiling range intermediate product (mainly C_{11}–C_{19} paraffins) could be 85.7% supposing only HDO reaction, and 80.8% supposing only decarboxylation/decarbonylation.) As at the favourable conversion (>95%), the ratio of the hydrocracking of the hydrocarbons was significant (>3%); so at the most favourable yield of INPRP was 80.5%. This approaches well the theoretical values. By increasing the temperature and the INPRP increasing effect of the pressure was lower and lower.

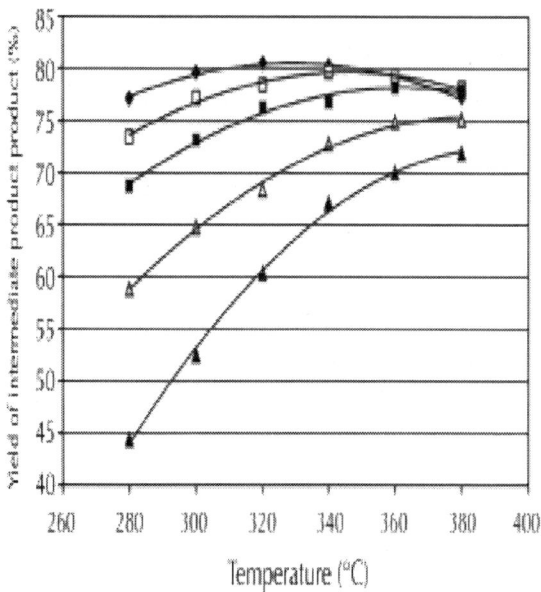

Figure 4: Yield of the INPRP as a function of the temperature and the LHSV; where:⬦-0.75 h−1, Δ-1.0 h−1, ⬦-1.5 h−1, □-2.0 h−1 and ■-3.0 h−1 of LHSV (P = 80 bar, H2/waste lard volume ratio = 600 Nm3/m3).

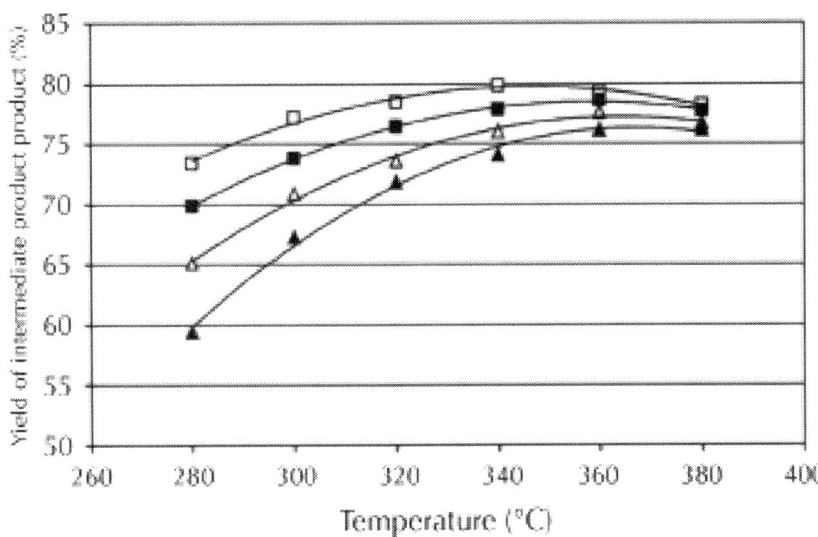

Figure 5: Yield of the INPRP as a function of the temperature and the pressure; where: ▲-20 bar, -40 bar, ■-60 bar and □-80 bar pressure (LHSV = 1.0 h⁻¹, H₂/waste lard volume ratio = 600 Nm³/m³).

To sum it up, we concluded that favourable yields of intermediate product were obtained at the following process parameters: T = 320–360°C, P = 40–80 bar, LHSV = 1.0–1.5 h⁻¹ H₂/feedstock ratio = 600 Nm³/m³. Application of the LHSV of 0.75 h⁻¹ is not only unadvised by the univocal decrease of the yield, but by the loss of plant capacity by 25% and 50% regarding the same rector capacity.

The INPRP contained olefins in a low concentration (iodine number between 1.0 and 10.0) at low temperature (≤320°C), low hydrogen pressure (≤40 bar) and high LHSV (≥1.5 h⁻¹), where the saturation of the olefinic double bonds was insufficient, but at the favourable process parameters, it was almost complete (iodine number < 1.0).

Oxygen Removing Reaction Paths and Their Ratio

During the conversion of triglycerides to the paraffin rich mixture, the saturation of double bonds, heteroatom removing (oxygen and other heteroatoms), isomerization and different side-reactions took place [8–

11]. The oxygen removal could happen by different reactions. In case of the hydrodeoxygenation (HDO) beside the normal paraffins having the same carbon number than that of the carboxyl acids composing the triglycerides only propane and water generate, while during the decarboxylation and decarbonylation the carbon number of the paraffins shortens by one and carbon-dioxide and carbon-monoxide also generate, respectively, beside the propane and water [12–15]. For example, Table 2 contains the composition of the gas products at 360°C, 80 bar, LHSV = 1.0 h⁻¹ and H_2/feedstock ratio = 600 Nm³/m³.

Table 2: Composition of the gas product (T = 360°C, P = 80 bar, LHSV = 1.0 h⁻¹, H_2/feedstock ratio = 600 Nm³/m³)

Component	Value	
	Mass %	Volume %
Hydrogen	23.94	85.86
Hydrogen-sulphide	0.45	0.09
Carbon-dioxide	28.39	4.66
Carbon-monoxide	7.18	1.85
Methane	1.30	0.59
Ethane	0.39	0.09
Propane	26.82	4.39
Iso-butane	0.11	0.01
Butane	0.17	0.02
Others	11.24	2.43

According to the composition of the INPRP, the ratio of the HDO to decarboxylation/decarbonylation reactions could be determined. But to determine the ratio of decarboxylation to decarbonylation the exact quantity of the CO, CO_2 and water has to be known, but that is not possible in the applied catalytic system, because of other side reactions (e.g., water-gas shift reaction, cracking of the hydrocarbons). The main part (>99.5%) of the carbon number of the fatty acids of the triglycerids of the used lard (feedstock) was an even number—similar to most of the natural triglycerides—wherein 98.0% was C_{16} and C_{18} carboxyl acids (Table1), so from the ratio of the generated C_{15} and C_{16} furthermore the C_{17} and C_{18} paraffins, the ratio of the hydrodeoxygenation (HDO) and the decarboxylation/decarbonylation reactions was determined. At

600 Nm³/m³ H₂/waste lard ratio and LHSV of 1.0 h⁻¹ which were found to be favourable furthermore in the whole investigated pressure range (see the corner points in Figure 6), the degree of the HDO and the other two reactions was almost the same (as C_{15} and C_{16} furthermore C_{17} and C_{18} paraffins were generated in similar quantity) at around 300°C. By increasing the temperature (320–360°C), the decarboxylation/decarbonylation reactions became dominant (the concentration of the odd carbon number paraffins was higher than that of the even carbon number paraffins).

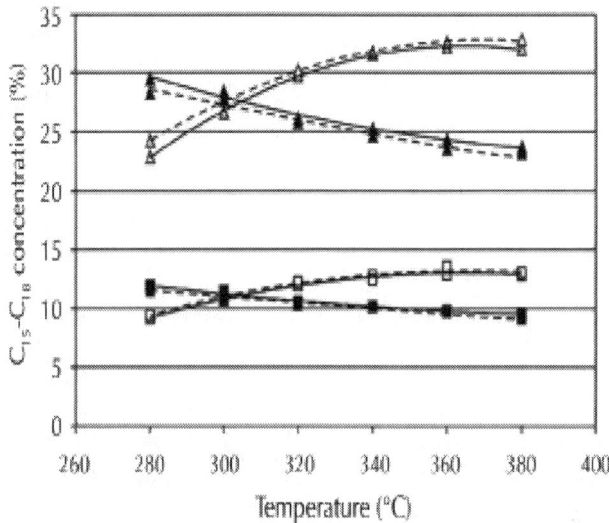

Figure 6: The concentration of the C_{15}–C_{18} paraffins of the intermediate product; where: □-C_{15}, ■-C_{16}, -C_{17} and ▲-C_{18} fraction (broken line: P = 20 bar, continuous line: P = 80 bar) (LHSV = 1.0 h⁻¹, H₂/feedstock volume ratio = 600 Nm³/m³).

The ratio of the hydrodeoxygenation (HDO) and decarboxylation/decarbonylation is presented as a function of the C_{18} and C_{17} paraffins (Figure 7). The ratio of the C_{18}/C_{17} paraffins decreased significantly by increasing the temperature and a little by decreasing the pressure over the applied catalyst in the investigated parameter range. This means that the ratio of decarboxylation/decarbonylation increased compared to the hydrodeoxygenation. Similar tendencies were found in case of the change of the ratio of the C_{15} and C_{16} paraffins.

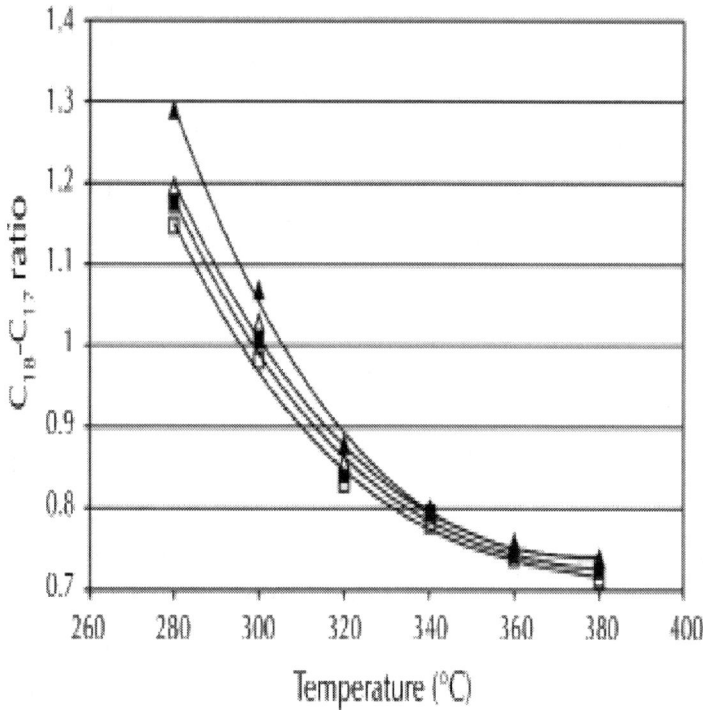

Figure 7: The ratio of C_{18}/C_{17} paraffins as a function of the temperature and the pressure; where: □-20 bar, ■-40 bar, -60 bar and ▲-80 bar pressure (LHSV = 1.0 h^{-1}, H$_2$/waste lard volume ratio = 600 Nm^3/m^3).

To determine the stability of the used catalyst and to produce a sufficient quantity of intermediate normal paraffin-rich product (INPRP) as the feedstock of the second stage of the bio gas oil production (isomerization) a 400-hour-long test was carried out at the favourable, expediently chosen parameter combination (T = 360°C, P = 60 bar; LHSV = 1.0 h^{-1}, H$_2$/waste lard volume ratio = 600 Nm^3/m^3). According to the results (Figure 8) the activity and the selectivity of the catalyst— after the initial decrease of activity—were unchanged during the experiment.

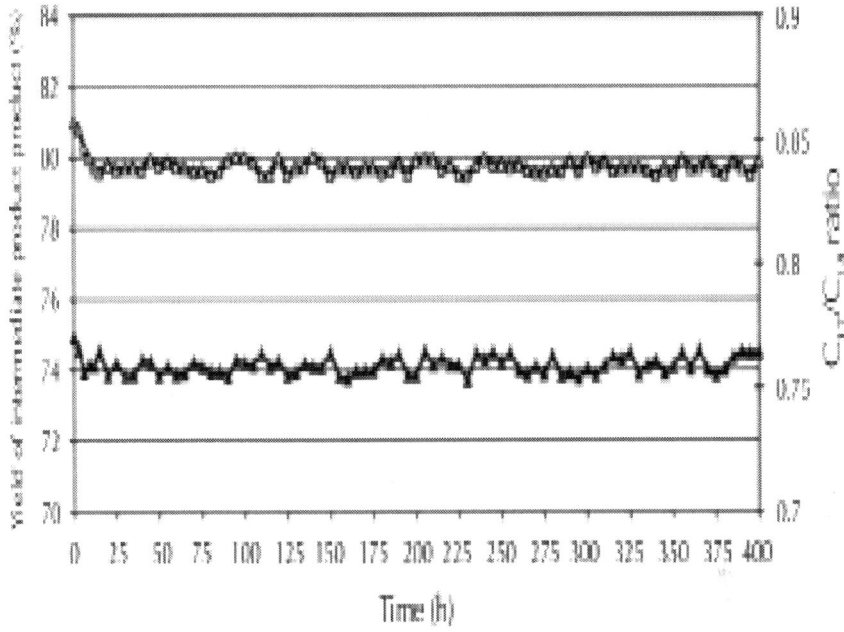

Figure 8: The change of the yield of the INPRP and the C_{17}/C_{18} ratio as a function of time; where: -yield of intermediate product and ▲-C_{17}/C_{18} ratio (T = 360°C, P = 60 bar, LHSV = 1.0 h⁻¹, H_2/waste lard volume ratio = 600 Nm³/m³).

The Service Properties of the Intermediate Product

The cetane number of the INPRP produced from waste lard over sulphided NiMo/Al_2O_3 catalyst was outstanding (101 unit) as it contained C_{15}–C_{18} paraffins in the highest degree (>95%) which have very high cetane number (100–105). Regarding the cold flow properties, the favourable fatty acid composition of this waste lard has to be highlighted. While the earlier investigated vegetable oils [13–15] contain mainly C_{18} or longer fatty acids (90%–95%), our lard contained 28.12% C_{16} fatty acids. Consequently, our products containing about 28% of C_{15} and C_{16} paraffins (of which freezing points are +10°C and +15°C, resp.) having more favourable CFPP (+19°C) than that of the product mixture (+23°C) [16] containing mainly (>90%) $C_{17}$and C_{18}

paraffins (freezing point +22°C and +28°C) produced over the same catalyst and at the same process parameters. Furthermore, as the isomers of the shorter carbon number paraffins have a lower freezing point than that of the longer ones, so by the isomerization of C_{15} and C_{16} paraffins a more favourable product could be obtained.

The Isomerization of the Normal Paraffin-Rich Intermediate Product

The catalytic isomerisation of the crude oil origin mixture having relatively high normal paraffin content (the other is cyclo and aromatic hydrocarbons) for improving the cold flow properties was carried out over noble metal containing catalysts [17–21]. Such catalysts with high isomerization activity are for example the different noble metal containing zeolites (ZSM-5, ZSM-22, ZSM-23), silica-alumina-phosphates (SAPO-11, SAPO-31, SAPO-41) and different mesoporous structures (MCM-41, Al-MCM-41) [17, 22]. But there are only a few articles dealing with the isomerisation of paraffin mixes of alternative source with different composition (e.g., containing oxygenic components) [4, 5, 16, 22].

During the isomerization experiment over Pt/SAPO-11, we used a mixture with high normal paraffin content produced from waste lard (Table 3). The composition of this feedstock and the products were investigated by GC-FID method using standard hydrocarbons. The feedstock was produced during the mentioned 400-hour-long experiment at process parameters (T = 360°C, P = 60 bar, LHSV = $1.0 \, h^{-1}$, H_2/lard ratio 600 Nm^3/m^3) that were found to be favourable for the hydro-deoxygenation of waste lard, because then-paraffin yield was high and the amount of the partially converted oxygen containing components (carboxylic-acids, esters, etc.) was very low in the product mixture. That was because the chosen Pt/SAPO-11 catalyst was found to be highly selective [5, 18–21] with great thermal stability [23]; however, it is highly sensitive for hydrolysis [16, 24]. According to the abovementioned, the investigation of the applicability of that is necessary in case of the feedstock produced from waste lard different from the already investigated ones.

Table 3: Main properties of the high n-paraffin containing feedstock

Properties		Value
Density at 40°C, g/cm3		0.7689
Cold filter plugging point, °C		21
Cetane number		101
Sulphur content, mg/kg		3.2
Nitrogen content, mg/kg		2.1
Paraffin content, %	i-C15	0.13
	n-C15	15.59
	i-C16	0.10
	n-C16	11.83
	i-C17	0.37
	n-C17	39.61
	i-C18	0.28
	n-C18	29.97
Total isoparaffin concentration %		0.92
Concentration of the oxygen containing compounds, %		0.43
Aromatic and cycloparaffin content, %		0.3

Yields of the Products

During our isomerisation experiment, we applied an H_2/intermediate product volume ratio found to be favourable during our previous experiments [16, 22], and we only examined the effects of the change of temperature, LHSV, and pressure which essentially affect the isomerization reactions. Consequently, the range of the applied process parameters was the following: temperature 320°C–380°C, total pressure 30–60 bar, liquid hourly space velocity (LHSV) = 1.0–3.0 h^{-1} and H_2/hydrocarbon volume ratio = 400 Nm³/m³.

The yield of the product mixtures decreased with increasing the temperature and with decreasing the pressure and the LHSV (Figures 9 and 10). The reason of this effect was that during the isomerization, the carbenium ions which have lower stability than the saturated hydrocarbons forming on the surface of the catalyst were cracked more easily at higher temperatures and at lower partial pressure caused by decreasing the hydrogen pressure, furthermore because of the lower LHSV and the longer contact time of the molecules on the surface of the catalyst. At the applied experimental process parameters, the yield of the products—expect for one case—exceeded 90%.

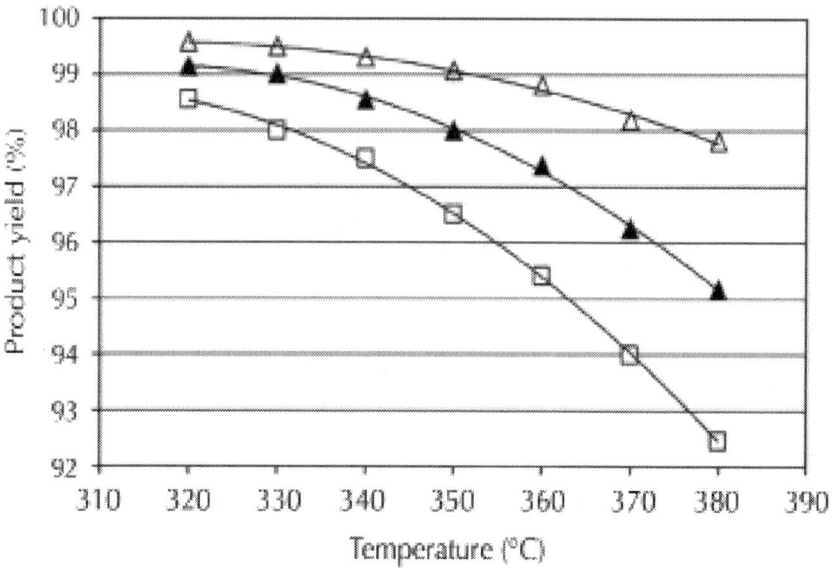

Figure 9: Product yields as a function of temperature and LHSV; where: □-1.0 h⁻¹, ▲-2.0 h⁻¹ and -3.0 h⁻¹ of LHSV (P = 40 bar, H_2/hydrocarbon volume ratio = 400 Nm³/m³).

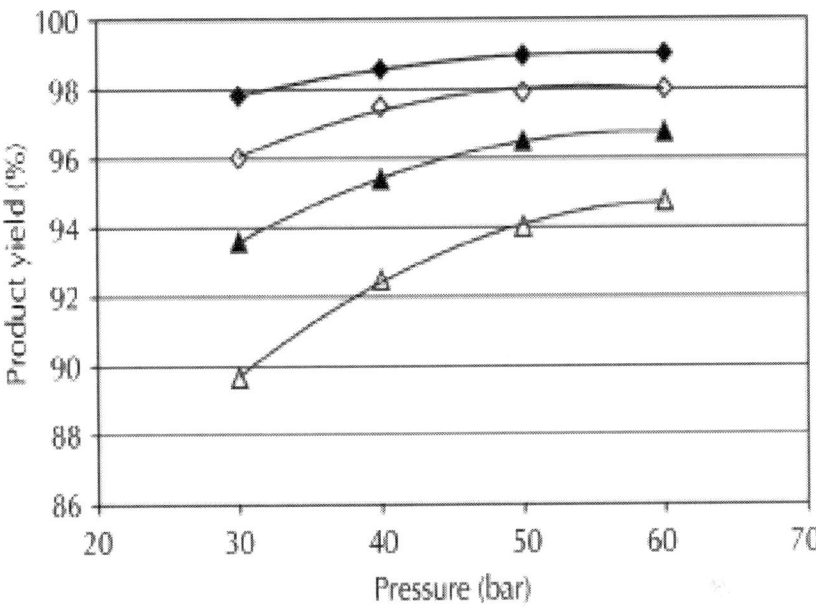

Figure 10: Product yields as a function of temperature and pressure; where ◆-320°C, ◊-340°C, ▲-360°C and -380°C (LHSV = 1.0 h⁻¹, H₂/ hydrocarbon volume ratio = 400 Nm³/m³).

The Composition of the Products

The isoparaffin content of the products significantly increased above 330°C with the increasing temperature, namely, the rate of isomerization increased. The degree of the increase in the isoparaffin yield at 360°C–370°C started to decrease, partly because of the closer approaching to the thermodynamic equilibrium, and partly because of the closer approach of the equilibrium concentrations and thermodynamic inhibition (the isomerization reactions are exoterm reactions) furthermore partly because of the intensifying cracking reactions. This tendency rose with decreasing the LHSV (Figure 11) as the contact time increased which called forth a higher rate of the cracking reactions.

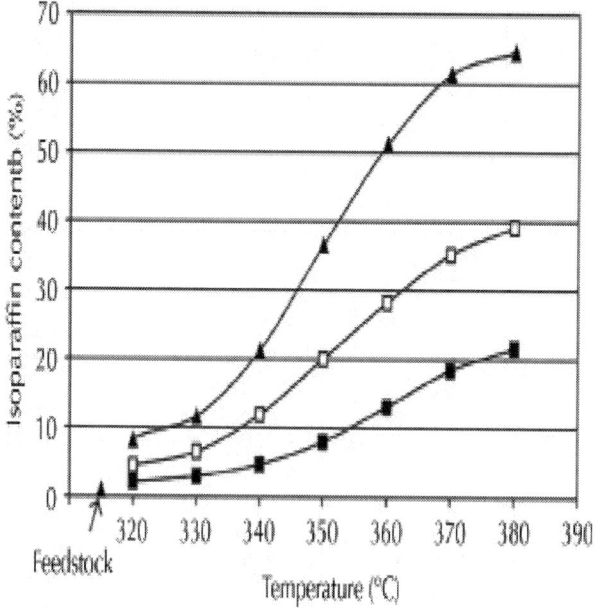

Figure 11: The isoparaffin content of the products as a function of temperature and LHSV; where: ▲-1.0 h⁻¹, □-2.0 h⁻¹ and ■-3.0 h⁻¹ of LHSV (P = 40 bar, H₂/hydrocarbon volume ratio = 400 Nm³/m³).

Out of the obtained isomers, up to 360°C, the product contained mainly (85%–90%) monobranched (mono-methyl) isomers of which freezing point is substantially lower (e.g., n-octadecane: +28°C; 2-methyl-heptadecane: +5,5°C, 5-methyl-tetradecane: −34,4°C) than that of the n-paraffins of the same carbon number and their cetane number is lower by only 20–25 unit than that of equivalent n-paraffins [25]. These latter values are at least 15–35 units above the 51 unit required by the European standard (EN 590:2009). However, above this process temperature multibranched isomers were also produced with increasing rate (15%–35%) of which cold flow properties are more favourable, but their cetane number is significantly lower [25]. So it is practical to determine a favourable compromise between the cold flow properties and the cetane number.

In the investigated range of process parameters, the decreasing pressure increased the rate of isomerization till 340°C. This occurred because by decreasing the partial pressure of the hydrogen, the first step of isomerization, namely, the dehydrogenation of hydrocarbons to olefins took place more easily on the active sites of the platinum catalyst. However, at 360°C the isoparaffin content reached a maximum as a function of the pressure, while at 380°C the decrease in the pressure caused a decrease in the isomer content (Figure 12). The reason of that was that the isoparaffin content was lowered by the intensive hydrocracking reactions at high temperatures, which rate was confined by the increase of the partial pressure of the hydrogen, namely, the rate of the hydrogenation of the labile carbenium ions generating on the surface of the catalyst increased.

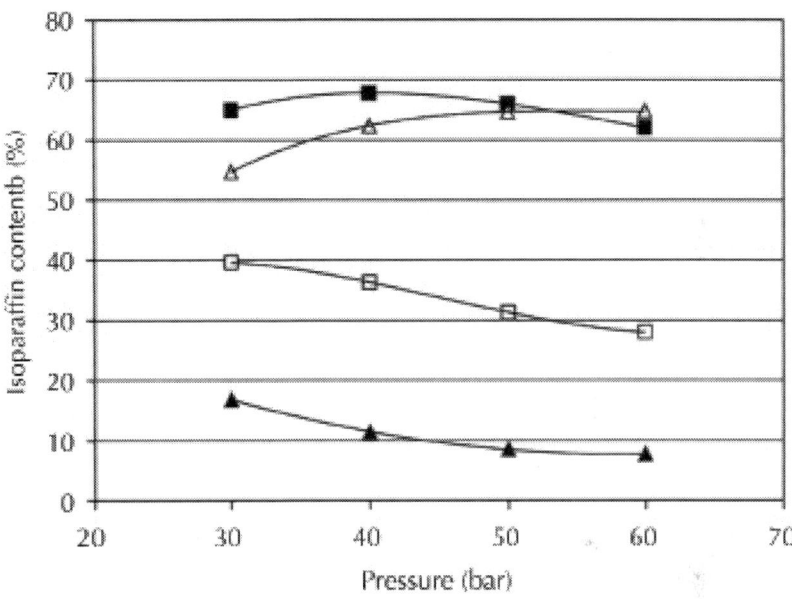

Figure 12: Total isoparaffin content of the products as a function of pressure, and temperature; where: ▲-320°C, □-340°C, ■-360°C and -380°C (LHSV = 1.0 h^{-1}, H$_2$/ hydrocarbon volume ratio = 400 Nm^3/ m^3).

The Main Properties of the Products

The cold filter plugging point (CFPP) is a very important property for diesel gas oils, because the paraffin crystals settled out with the decreasing temperature could cause plugging in the fuel filter and this could cause unserviceability in the fuel supply system. The CFPP values of the products decreased by increasing the temperature and by decreasing the pressure and the LHSV (Figure 13). The cause of this effect was mainly the increase in the isoparaffin content of the product, because their freezing-point is significantly lower than that of the n-paraffins. Besides the abovementioned, the lower freezing-point light hydrocarbons forming in the hydrocracking reactions also helped to improve the CFPP values of the products.

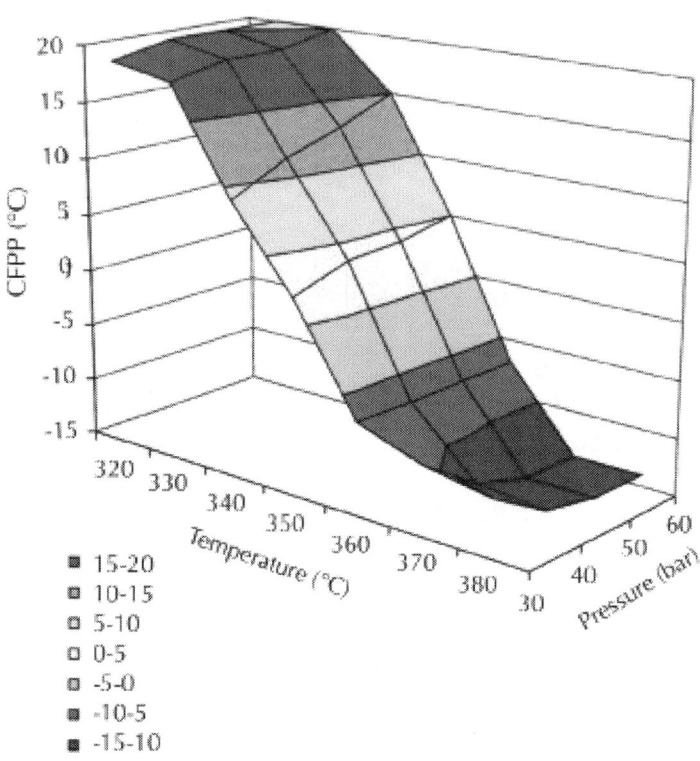

Figure 13: CFFP values of the products as a function of temperature, pressure and LHSV.

The increasing temperature (in the investigated range) positively affected the CFPP values (unequivocally lowering it). The reason of this was that besides the mono-branched isomers, the concentration of the multi-branched isomers also increased, because at higher temperatures not only the reaction rate of the forming of the mono-branched isomers increased, but in consecutive reactions there is a potential of the forming of multi-branched isomers, as well. These components have a positive effect on the cold flow properties; however, their other important property, the cetane number, is by far more unfavourable. Therefore the cetane number of these products decreased compared to the normal paraffin mixture (101 unit), because of the increasingly produced multi-branched isomers. However, the products obtained at the favourable process parameters (T = 360–370°C, P = 50 bar, LHSV = 1.0 h^{-1}, H$_2$/intermediate product volume ratio = 400 Nm^3/m^3) determined by compromises (yield of organic liquid phase, ratio of multi-branched i-paraffins, location of the branches) had cetane number between 76–88, which greatly exceeds the 51 specified in the valid standard, while their CFPP values were between −5 and −15°C.

CONCLUSIONS

Based on our experimental results it was concluded that on the expediently chosen sulphided NiMo/Al$_2$O$_3$ catalyst and at the favourable process parameters (T = 320–360°C, P = 40–80 bar, LHSV = 0.75–1.5 h^{-1}, H$_2$/waste lard volume ratio = 600 Nm^3/m^3) the INPRP having gas oil boiling range contained merely paraffins (>99.5%) which were mainly C$_{15}$, C$_{16}$, C$_{17}$ and C$_{18}$ paraffins, namely on the applied catalyst both the HDO and the decarboxylation/decarbonylation reactions took place, as well. The yields of the intermediate product were high (73.9%–79.7%), which approaches well the theoretical values that could be reached in case of the HDO and decarboxylation/decarbonylation reactions; 80.8 and 85.7, resp.). At 300°C, the two main reaction pathways took place at a nearly equivalent degree; by increasing the temperature (320–360°C) the decarboxylation/decarbonylation reactions came to the front. The cold flow properties (e.g., cold filter plugging point) of the intermediate products were unfavourable with 19°C, which was decreased to between −5 and −15 by the isomerisation over 0.5% Pt/SAPO-11 (T = 360–370°C, P = 50 bar, LHSV = 1.0 h^{-1}, H$_2$/intermediate product volume ratio = 400 Nm^3/m^3). The obtained products are excellent waste-origin blending components for diesel gas oils as their cetane number is high (76–88), and because these products are also aromatic-free, burn cleanly, and are environmentally friendly.

REFERENCES

1. A. Demirbas, "Biorefineries: current activities and future developments," Energy Conversion and Management, vol. 50, no. 11, pp. 2782–2801, 2009.

2. M. Balat and H. Balat, "A critical review of bio-diesel as a vehicular fuel," Energy Conversion and Management, vol. 49, no. 10, pp. 2727–2741, 2008.

3. S. Jain and M. P. Sharma, "Stability of biodiesel and its blends: a review," Renewable and Sustainable Energy Reviews, vol. 14, no. 2, pp. 667–678, 2010.

4. J. Hancsók, M. Krár, S. Magyar, L. Boda, A. Holló, and D. Kalló, "Investigation of the production of high cetane number bio gas oil from pre-hydrogenated vegetable oils over Pt/HZSM-22/AlO,"Microporous and Mesoporous Materials, vol. 101, no. 1-2, pp. 148–152, 2007.

5. J. Hancsók, M. Krár, SZ. Magyar, L. Boda, A. Holló, and D. Kalló, "Investigation of the production of high quality biogasoil from pre-hydrogenatedvegetable oils over Pt/SAPO-11/AlO," Studies in Surface Science and Catalysis, vol. 170, pp. 1605–1610, 2007

6. J. Gergely, J. Petro, G. Resofszki, et al., "Production of selective hydroisomerization catalyst and selective hydroisomerization of hydrocarbons," Hung. Pat. 225 912.

7. J. Hancsók, G. Gárdos, and M. Baumann, "The effect of platinum dispersion on the isomerisation of n-pentane," Hungarian Journal of Industrial Chemistry, vol. 17, pp. 131–137, 1989.

8. B. Donnis, R. G. Egeberg, P. Blom, and K. G. Knudsen, "Hydroprocessing of bio-oils and oxygenates to hydrocarbons. Understanding the reaction routes," Topics in Catalysis, vol. 52, no. 3, pp. 229–240, 2009.

9. G. W. Huber and A. Corma, "Synergies between bio- and oil refineries for the production of fuels from biomass," Angewandte Chemie International Edition, vol. 46, no. 38, pp. 7184–7201, 2007

10. G. N. da Rocha Filho, D. Brodzki, and G. Djéga-Mariadassou, "Formation of alkanes, alkylcycloalkanes and alkylbenzenes during the catalytic hydrocracking of vegetable oils," Fuel, vol.

72, no. 4, pp. 543–549, 1993.

11. S. Bezergianni and A. Kalogianni, "Hydrocracking of used cooking oil for biofuels production,"Bioresource Technology, vol. 100, no. 17, pp. 3927–3932, 2009.

12. J. Gusmão, D. Brodzki, G. Djéga-Mariadassou, and R. Frety, "Utilization of vegetable oils as an alternative source for diesel-type fuel: hydrocracking on reduced Ni/SiO_2 and sulphided Ni-Mo/ -Al_2O_3," Catalysis Today, vol. 5, no. 4, pp. 533–544, 1989.

13. M. Krár, A. Thernesz, Cs. Tóth, T. Kasza, and J. Hancsók, Silica and Silicates in Modern Catalysis, Transworld Research Network, Kerala, India, 2010.

14. P. Šimá ek, D. Kubi ka, G. Šebor, and M. Pospíšil, "Hydroprocessed rapeseed oil as a source of hydrocarbon-based biodiesel," Fuel, vol. 88, no. 3, pp. 456–460, 2009.

15. I. Kubi ková, M. Snåre, K. Eränen, P. Mäki-Arvela, and D. Y. Murzin, "Hydrocarbons for diesel fuel via decarboxylation of vegetable oils," Catalysis Today, vol. 106, no. 1-4, pp. 197–200, 2005.

16. T. Kasza, A. Holló, A. Thernesz, and J. Hancsók, "Production of bio gas oil from bioparaffins over Pt/SAPO-11," Chemical Engineering Transactions, vol. 21, pp. 1225–1230, 2010.

17. C. Perego, "Biomass to fuels: the challenges of zeolite and mesoporous materials," in Proceedings of the 16th International Zeolite Conference Joint with the 7th International Mesostructureed Materials Symposium, Sorrento, Italy, July 2010, KN-8.

18. R. J. Taylor and R. H. Petty, "Selective hydroisomerization of long chain normal paraffins," Applied Catalysis A, vol. 119, no. 1, pp. 121–138, 1994.

19. T. Blasco, A. Chica, A. Corma, W. J. Murphy, J. Agúndez-Rodríguez, and J. Pérez-Pariente, "Changing the Si distribution in SAPO-11 by synthesis with surfactants improves the hydroisomerization/dewaxing properties," Journal of Catalysis, vol. 242, no. 1, pp. 153–161, 2006

20. J. Walendziewski and B. Pniak, "Synthesis, physicochemical properties and hydroisomerization activity of SAPO-11 based catalysts," Applied Catalysis A, vol. 250, no. 1, pp. 39–47, 2003.

21. J. M. Campelo, F. Lafont, and J. M. Marinas, "Hydroconversion of n-dodecane over Pt/SAPO-11 catalyst," Applied Catalysis A, vol. 170, no. 1, pp. 139–144, 1998.

22. J. Hancsók, S. Kovács, Gy. Pölczmann, and T. Kasza, "Investigation of the effect of oxygenic compounds on the isomerization of bioparaffins over Pt/SAPO-11 catalyst," in Proceedings of the 14th Nordic Symposium on Catalysis, Marienlyst, Denmark, August 2010, Book of Abstracts P26.

23. A. Buchholz, W. Wang, M. Xu, A. Arnold, and M. Hunger, "Thermal stability and dehydroxylation of brønsted acid sites in silicoaluminophosphates H-SAPO-11, H-SAPO-18, H-SAPO-31, and H-SAPO-34 investigated by multi-nuclear solid-state NMR spectroscopy," Microporous and Mesoporous Materials, vol. 56, no. 3, pp. 267–278, 2002.

24. W. Lutz, R. Kurzhals, S. Sauerbeck et al., "Hydrothermal stability of zeolite SAPO-11," Microporous and Mesoporous Materials, vol. 132, no. 1-2, pp. 31–36, 2010.

25. R. C. Santana, P. T. Do, M. Santikunaporn et al., "Evaluation of different reaction strategies for the improvement of cetane number in diesel fuels," Fuel, vol. 85, no. 5-6, pp. 643–656, 2006.

Chapter 2

Oil Well Characterization and Artificial Gas Lift Optimization Using Neural Networks Combined with Genetic Algorithm

Chukwuka G. Monyei[1], Aderemi O. Adewumi[2], and
Michael O. Obolo[3]

[1]Department of Electrical and Electronic Engineering, University of Ibadan, Nigeria

[2]School of Mathematics, Statistics and Computer Science, University of KwaZulu-Natal, South Africa

[3]Department of Petroleum Engineering, University of Ibadan, Nigeria

ABSTRACT

This paper examines the characterization of six oil wells and the allocation of gas considering limited and unlimited case scenario. Artificial gas lift involves injecting high-pressured gas from the surface into the producing fluid column through one or more subsurface valves set at predetermined depths. This improves recovery by reducing the bottom-hole pressure at which wells become uneconomical and are thus abandoned. This paper presents a successive application of modified artificial neural network (MANN) combined with a mild intrusive genetic algorithm (MIGA) to the oil well characteristics with promising results This method helps to prevent the overallocation of gas to wells for recovery purposes while also maximizing oil production by ensuring that computed allocation configuration ensures maximum economic accrual. Results obtained show marked improvements in the allocation especially in terms of economic returns.

INTRODUCTION

Petroleum, a limited natural resource, is a nonrenewable form of energy on which humans largely depend. This leads to pressing market demands, accessibility issues, and competitive market environment that force oil companies to seek technologies and procedures that can give competitive advantage and meet environmental restrictions while streamlining production processes and cutting costs [1].

Artificial gas lift (AGL) is a recovery process that involves the use of gases, produced (from oil) or purchased, which are pumped into the well bore to maintain formation pressure, that is, the pressure at which the fluid flows to the surface. There are two types of AGL, namely, intermittent gas lift and continuous gas lift [1]. However, this paper is not concerned with the types of AGL but rather its distribution or allocation.

The gas lift process involves the injection of high pressure gas at the bottom of the production tubing of an oil well [1–3]. In other words, AGL involves injecting high-pressured gas from the surface into the producing fluid column through one or more subsurface valves set at predetermined depths [2, 3]. This helps to improve recovery by reducing the bottom-hole pressure at which wells become uneconomic, resulting in being abandoned. The gas, mixed with the oil, diminishes the weight of the fluid column thereby reducing the downhole pressure. A low downhole pressure induces a flux of fluids from the reservoir to the well. The produced fluid is composed of oil, gas, and water. The water must be treated before being discharged which incurs costs while the gas can be either reused in the process or sent to customers and other facilities [3]. In large oil fields, several separators are used to divide the three phases. This gives rise to the problem of maximizing production by allocating lift-gas to the wells while defining the routing from wells to separators and observing separator capacities [2, 3].

In particular, the gas lift operation of oil fields is one of many production processes whose performances can be improved. As the internal pressure in high-depth or depleted reservoirs can force the flow of only a fraction of their oil to the surface, the use of artificial means becomes imperative to lift the oil, especially for deep reservoirs that are found off-shore. Two examples of artificial lifting are submerged pumps and continuous injection of gas [3]. Although the former can, in principle, recover most of the oil, its operating costs are excessively high for today's oil prices, not to mention the potential of an unfavourable energy trade and other technical hindrances. The gas lift technique, on the other hand, harnesses the reservoir's gas by injecting natural gas into the production tubing so as to reduce the weight of the oil column, thereby elevating the mix of oil, gas, and water to the surface.

A motivation for this work therefore arises from the need to reduce wastage in gas allocation especially in the unlimited scenario thus freeing up gas thus freeing up gas for other uses such as domestic, transportation and electricity generation purposes. With global outcry to the insidious effect of green-house gases on our environment and the need for prudent management of scarce resources, this paper aims at providing a cost-effective method for solving the problem of gas allocation for recovery purposes in the oil and gas industry.

The rest of this paper is organized as follows: Section 2 presents a brief overview of related works in literature while Section 3 further describes the problem as well as modelling approach. Section 4 gives an overview of the methodology adopted to solve the oil well characterization and AGL problem while results obtained are discussed in Section 5. The final section presents some useful conclusions as well as direction for further works.

LITERATURE REVIEW

Gas injection has been used to maintain reservoir pressure at some selected levels or to supplement natural reservoir energy to a lesser degree by reinjection of a portion of the produced gas. Complete or partial pressure maintenance operations can result in increased hydrocarbon recovery and improved reservoir production characteristics. A general position opines that daily oil production increases concomitantly with gas up to a certain level where further gas injection yields a decrease in oil production with increased gas cost [2]. Ray and Sarker [2] developed a multiobjective constrained algorithm to optimize gas lift allocation within the constraint of limited available gas. The proposed solution was applied to six and fifty-six well problems with single and multiobjective problem formulations. De Souza et al. [4] described a case study that involved modelling and optimizing gas allocation for deep water offshore petroleum production with interest in determining the rate of injected gas flow that guarantees maximum oil production, profit, and optimal design of gas lift system considering capital cost of compressors, turbine, and gas pipeline constraints. The problem was modelled as a nonlinear optimization problem and solved as a two-phase network flow model. Codas and Camponogara [1] addressed the problem of gas lift allocation with separator routing constraint using a mixed integer linear model solved using the CPLEX software. Mahmudi and Sadeghi [5] used a hybrid computational model consisting of genetic algorithm (GA) and Marquardt algorithm to optimize gas allocation under various constraints including effects of tubing diameter, rates of gas injection, and separator pressure on the economic return of the well over a long period.

From the foregone and other research trends, the problem of well characterization and gas allocation is mainly addressed independently. Moreover, specific interests of earlier works did not focus on configuration selection but rather on routing mechanisms such as separator routing constraint, effects of tubing diameter, rates of gas injection and separator pressure on the economic return of the well over long period, and capital cost of compressors, turbine, and gas pipeline amongst others. Most previous works therefore focused on the contribution of the artificial gas lift layout and material selection in ensuring optimum gas yield. This paper seeks to complement ongoing research by proposing a combined modified artificial neural network (MANN) and mild intrusive genetic algorithm (MIGA) intelligent technique for optimum well characterization and gas lift allocation to achieve maximum economic yield. We seek to characterize the oil wells and get maximum produced oil using limited available gas which indirectly results in increased economic returns. Further details and description of the gas lift problem can be found in [2–4, 6–8].

PROBLEM DEFINITION

Statement of the Problem

An oil field consisting of six wells is considered in this paper. AGL is employed in improving the recovery of the respective wells. Available gases at times are not sufficient to guarantee maximum oil production from the wells while classical techniques also do not guarantee optimal allocation of these available gases; hence an efficient algorithm that can optimally allocate them by selecting the best configuration that can ensure optimum economic accrual is sought. This paper proposes the MANN-MIGA approach for this problem.

Modelling

This paper addresses a case scenario involving six wells and limited gas supply that is currently not able to guarantee maximum production. We employ modified versions of artificial neural networks (ANN) and GA which has been successfully used in literature (e.g., see [9, 10]). These algorithms seek to optimize the allocation of available gas quantity among the six wells under consideration by allocating quantities that guarantee maximum economic accrual. First, MANN is used to characterise the oil production in terms of B/D (B/D refers to unit of measuring oil production output in terms of barrels per day) and the capacity of each well with respect to the gas injected (in MMscf/D), MMscf/D refers to unit of measuring gases in terms of million standard cubic feet per day. Next, MIGA is used to select the best values for each oil well while observing the following.

- Total gas allocated for the six wells does not exceed the available gas quantity for optimization.
- Economic accrual is more important than allocating entire gas and must therefore be maximized.
- Where there are two feasible values for gas allocation with the same oil production, the lowest value is chosen.

The work seeks to address the issue of efficient usage of scarce gas thus preventing wastage and enhancing oil production. In the event of other uses of gas emanating, optimization of the available gas for optimum oil production, maximum economic accrual, and other emanating needs is possible.

The main objective is to develop a strategy for the optimal allocation of limited (and unlimited) gas supply in an oil field involving six wells. Gas allocation is to be minimized while profit from both the sale of oil and remaining unallocated gas is to be maximized by maximizing oil production. We assume full dispatch of gas thus neglecting interaction between gas, conveying media, and other intermeddling media.

If i represents the index of the wells and n is the number of wells, the objective function is formulated as follows

$$\text{Minimize gas allocation} = \sum_{i=1}^{n} [G_u(\cdot)],$$

$$\text{Maximize oil production} = \sum_{i=1}^{n} [O_u(\cdot)]$$

$$= \sum_{i=1}^{n} \left[a + bG_u(\cdot) \right.$$

$$\left. + cG_u(\cdot)^2 + dG_u(\cdot)^3 \right],$$

$$\text{Maximize profit from sales} = P_g \sum_{i=1}^{n} [O_u(\cdot)] + P_o O_m,$$

$$(1)$$

where $G_u(.)$ is gas unit allocated for well I; $O_u(.)$ is respective oil produced for respective gas allocated to well i; a, b, c, and d are constants for derived polynomial function and P_g is unit price of gas ($/B); P_o is unit price of oil ($/MMscf); O_m is unallocated gas $= T_a - \sum_{i=1}^{n} [G_u(.)]$; T_a is (un)limited gas available for allocation.

Subject to
(1)

$$\sum_{i=1}^{n} [G_u (\cdot)] \leq T_a \gamma i;$$

(2)

(2) i, if G_i is a or b and $G_i(a)=G(b)$, where O_i is oil produced for $G_i(.)$, if .a<b gas allocated=$G_i(a)$

Also, the equation governing gas allocation (MMscf/D) and oil produced (B/D) is given as

$$Q_{Li} = aQ_{Qi}^2 + bQ_{Qi} + c,$$

(3)

where i is well number, Q_{Li} is oil produced per well (in B/D), Q_{Qi} is gas allocated (in MMscf/D), and a, b, and care constants

The value of the constants a, b, and c for the six wells is given in Table 1.

Table 1

Well	Value a	Value b	Value c
1	1138.71	799.94	−28.00
2	841.34	893.50	−277.69
3	131.28	61.54	0
4	135.92	39.08	0
5	125.38	49.05	0
6	156.35	89.46	0

The values used in training the MANN (Section 4) were obtained from [8].

METHODOLOGY

As stated earlier, this work adopts combined algorithms of MANN-MIGA which are essential modified forms of ANN and GA. The choice lies in the facts that the underlying techniques have been successfully used in other similar problems as they offer needed efficiency, speed, and flexibility [6, 7]. Moreover, the combination of MANN and MIGA has been efficiently used to solve other problems in literature with promising results [9, 10]. We present below a detailed description of the methods as applied to the current problem.

Modified Artificial Neural Network

In characterizing the gas injection/oil production of the wells, MANN was used in generating a model. In designing the algorithm for the neural network, a simple regression formula was used. The algorithm receives the inputs, sorts them out, adjusts its parameters, and computes the expected result. A step by step overview of the algorithm is presented as follows.

- The inputs are received in the proper (same) dimension (Figure 1).

- Interpolation is then carried out on each input matrix and each output matrix using their respective minimum and maximum values in generating an equivalent value of the contribution of each input to the expected answer or result.

- Received interpolated input(s) and output(s) are sorted (ranked) concurrently in order to obtain a minimum and maximum value with the first matrix serving as a baseline (Figure 2).

- Since MATLAB makes use of matrices, received inputs are thus stored in matrices. The output matrix is checked for crossovers (transition from a high-low-high or low-high-low number) using the first input matrix as baseline.

- The transitions obtained from the output matrix are used in generating the order of the polynomial function or curve in which our data is to be fitted into (5)–(10). As can be surmised from (5)–(10), there are two inputs and one output (see Figure 1). The modelling of the first input is calculated using (5)–(7) which are used in generating the respective values of a, b, and c. Our first output is thus given as $K(I)=aI^2+bI=c$. The second input is modeled similarly using (8)–(10) which are used in generating the respective values of a1, b1 and c1. The second output is thus given as $K(J)=a1J^2+b1J+c1$. Our final output is thus the average of these two values given as.$K=(K(I)+K(J))/2$.

- Generated value(s) is/are then recalculated in order to obtain actual values which are then fed forward to the output.

- Before the generated values are displayed, they are adjusted for errors using appropriate weights. For the purpose of this analysis, the weights have been assumed to be unity and there is no back propagation network provided to assist in adjusting the displayed values. The inputs therefore must be reliable and fairly accurate.

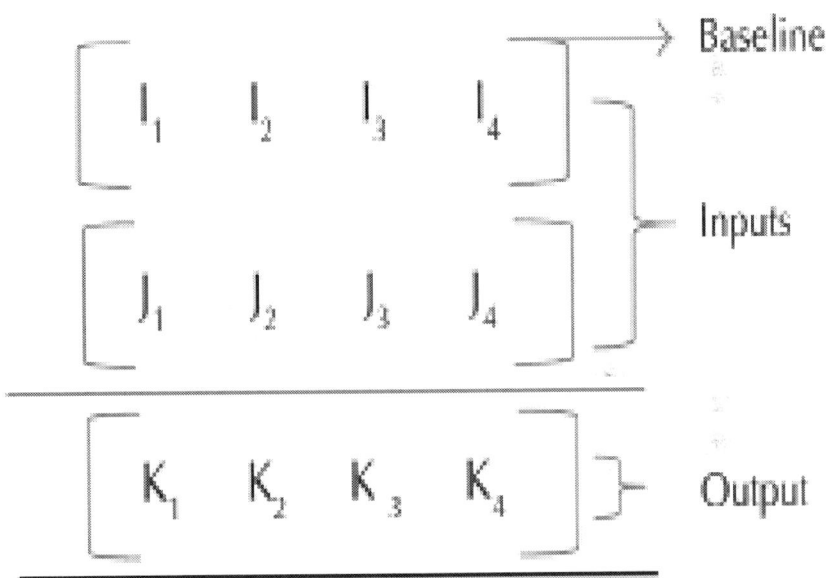

Figure 1: Received inputs and output during training.

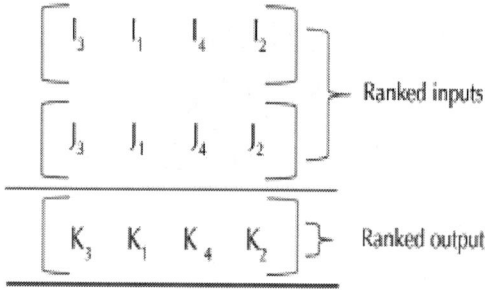

Figure 2: Ranked received inputs and output.

If

$$K_1 > K_3$$

$$K_1 > K_4$$

$$K_2 < K_4,$$

(4)

where I_i, J_i, and K_i (i=1,2,3,4) are valid inputs and output, respectively, then the number of crossovers could be determined from the ranked output. From (1) therefore, the ranked output of Figure 2 has 1 crossover.

One crossover therefore signifies that our data could be modeled using a polynomial of order 2 (i.e., a quadratic equation). In modelling therefore the following equations are used:

$$\sum K_i = a * n + b\sum_{i=1}^{n} I_i + c\sum_{i=1}^{n} I_i^2,$$

(5)

$$\sum K_i I_i = a\sum_{i=1}^{n} I_i + b\sum_{i=1}^{n} I_i^2 + c\sum_{i=1}^{n} I_i^3,$$

(6)

$$\sum K_i I_i^2 = a \sum_{i=1}^{n} I_i^2 + b \sum_{i=1}^{n} I_i^3 + c \sum_{i=1}^{n} I_i^4,$$ (7)

$$\sum K_i = a1 * n + b1 \sum_{i=1}^{n} J_i + c1 \sum_{i=1}^{n} J_i^2,$$ (8)

$$\sum K_i J_i = a1 \sum_{i=1}^{n} J_i + b1 \sum_{i=1}^{n} J_i^2 + c1 \sum_{i=1}^{n} J_i^3,$$ (9)

$$\sum K_i J_i^2 = a1 \sum_{i=1}^{n} J_i^2 + b1 \sum_{i=1}^{n} J_i^3 + c1 \sum_{i=1}^{n} J_i^4.$$ (10)

Mild Intrusive Genetic Algorithm

MIGA is used in combination with the MANN in arriving at optimum solutions for gas allocation under limited and unlimited conditions. GA is an evolutionary algorithm that mimics the principle of natural selection, reproduction, and survival of the fittest in solving complex optimization problems [11]. It has been widely and successfully used in literature howbeit in modified or improved form from the standard GA [11, 12] when it comes to some complex problems [13–15]. GA has also been applied to difficult problem involving the control of gas pipeline transmission [6, 7].

MIGA is designed as a modified population-based technique. Figure 3 provides a sample space that illustrates the environment for the activities of MIGA. GA allows for various ways of defining the population structure (chromosome) [14, 16].

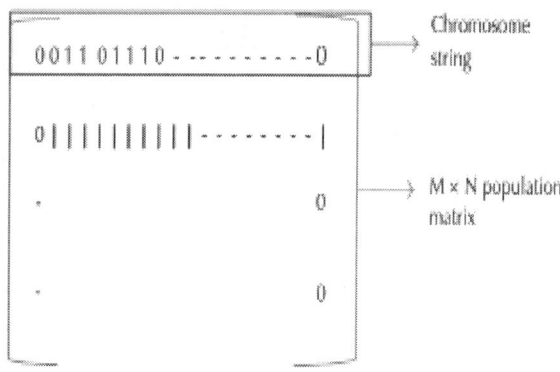

Figure 3: Sample space for MIGA.

In this study, MIGA uses the binary representation. It then runs through the usual steps of selection, crossover, elitism, and mutation as illustrated in Figure 5. Each chromosome string (Figure 4) corresponds to a solution whose fitness is tested for optimality (fitness value of zero). Figure 5(b) shows an instance of a poor solution with fitness value farther from zero.

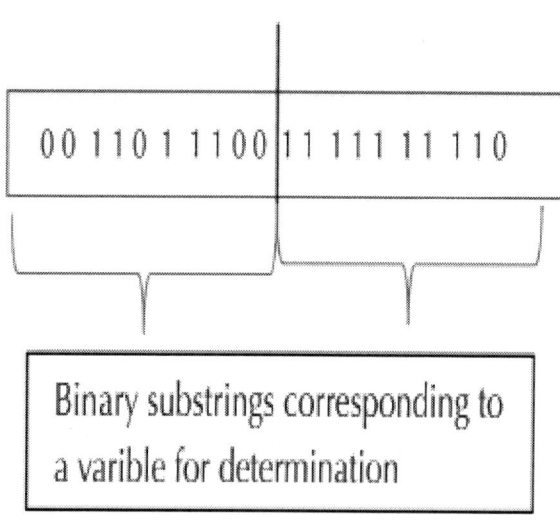

Figure 4

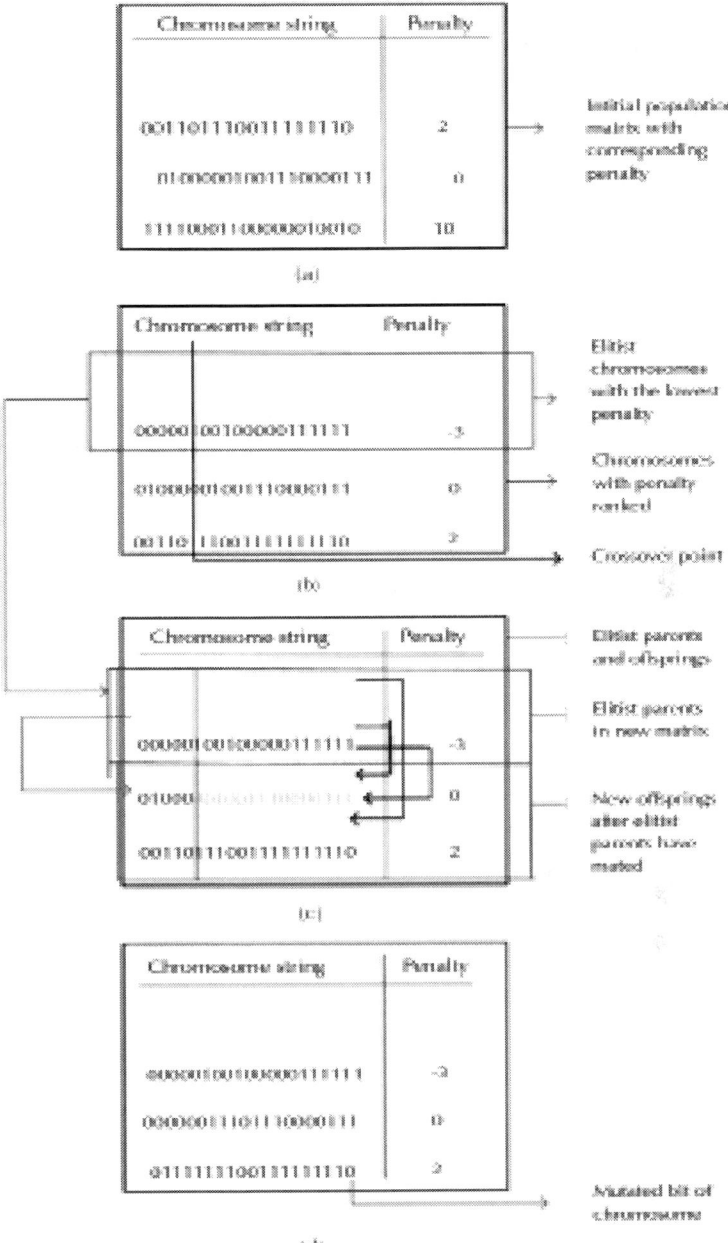

Figure 5: Illustration of MIGA steps.

SIMULATION EXPERIMENT, RESULTS, AND DISCUSSIONS

Simulations Environments

Simulation experiment for this work was done with MATLAB R2009a. Results were interpreted in form of graphs (as shown in Figures 6–13) and tables (as depicted in Tables 2 and 3) generated within this environment. The figures are grouped sometimes into three groups based on the well allocation number as presented and briefly discussed below. Experiment was done on a system with 4 GB DDR3 Memory, 500 GB HDD, and an Intel Core™ i3-380 M Processor.

Table 2: Limited gas optimization values for different well combinations

Well number	1	1-2	1–3	1–6
Optimum value (MMscf/D)	1.02	2.16	3.29	7.71
Given value (MMscf/D)	1	2	3	5
Allocated (MMscf/D)	Well 1: 0.999	Well 1: 0.9666	Well 1: 0.96933	Well 1: 0.75867
		Well 2: 1.0324	Well 2: 1.0285	Well 2: 0.78643
			Well 3: 1.0018	Well 3: 0.75565
				Well 4: 0.50765
				Well 5: 1.2298
				Well 6: 0.96
Unallocated (MMscf/D)	0.001	0.001	0.00037	0.0018

% allocation	99.9	99.9	99.99	99.96

Table 3: Unlimited gas optimization values for different well combinations

Well number	1	1-2	1–3	1–6
Optimum value (MMscf/D)	1.02	2.16	3.29	7.71
Given value (MMscf/D)	1.5	3	5	9
Allocated (MMscf/D)	Well 1: 1.02	Well 1: 1.0173 Well 2: 1.14	Well 1: 0.996 Well 2: 1.14 Well 3: 1.13	Well 1: 0.81467 Well 2: 1.0593 Well 3: 1.0592 Well 4: 1.99 Well 5: 1.4112 Well 6: 0.99554
Unallocated (MMscf/D)	nil	0.0027	0.024	1.38
% allocation	100	99.88	99.27	82.1

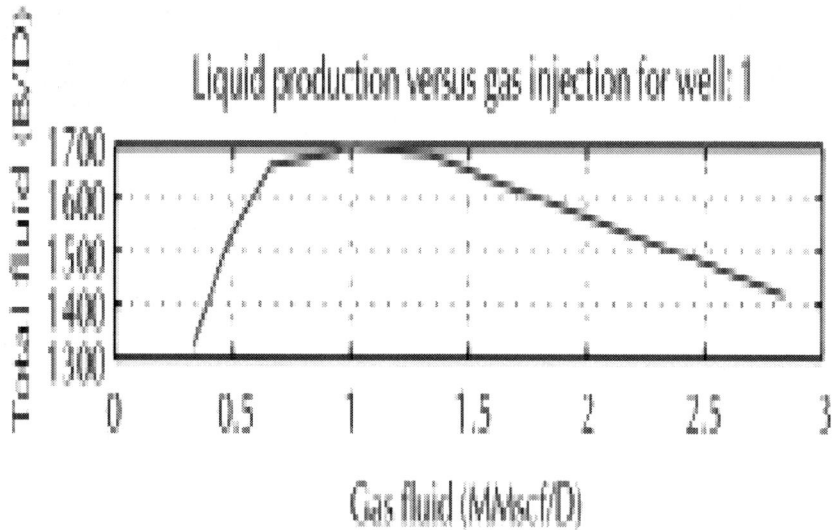

(a)

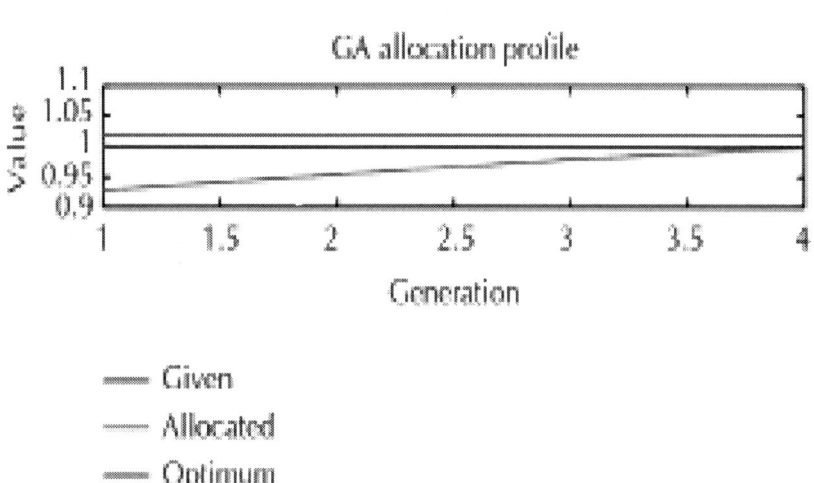

(b)

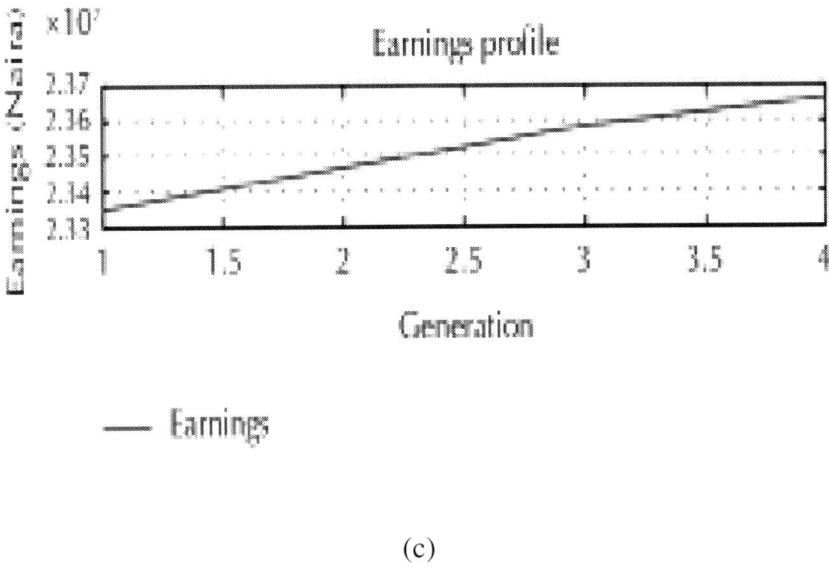

(c)

Figure 6: Well 1's allocation profile, GA allocation profile, and the earnings (Naira) profile for a limited scenario.

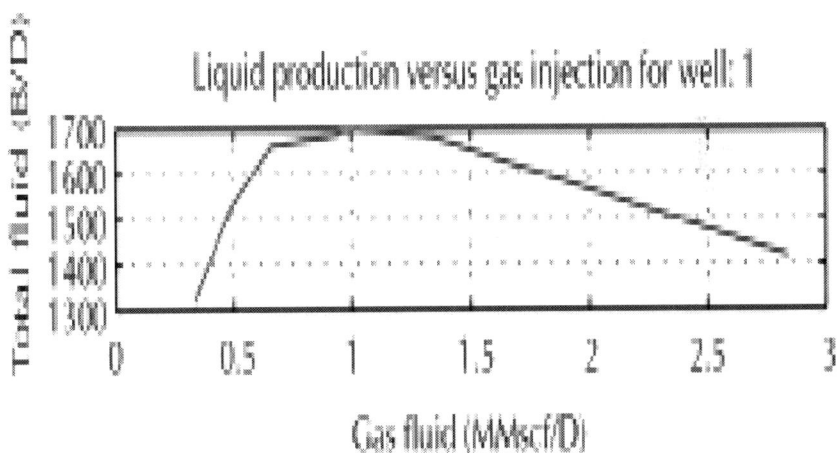

(a)

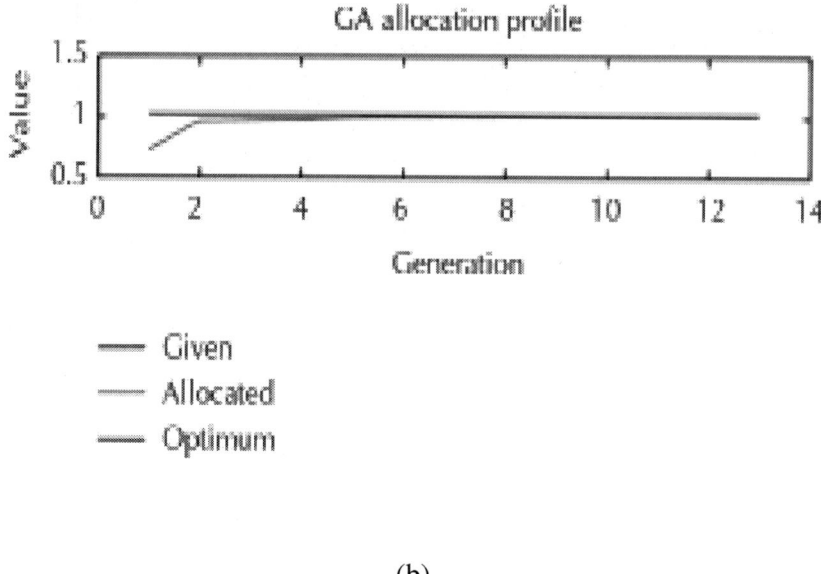

(b)

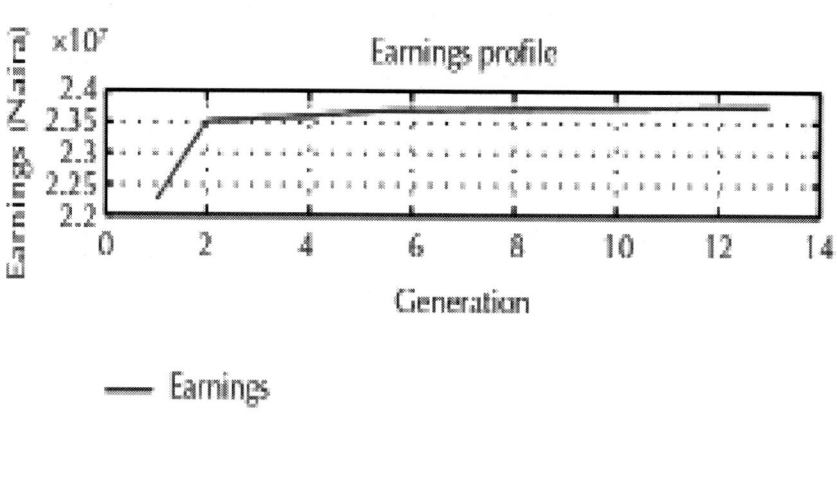

(c)

Figure 7: Well 1's allocation profile, GA allocation profile, and earnings (Naira) profile in an unlimited scenario.

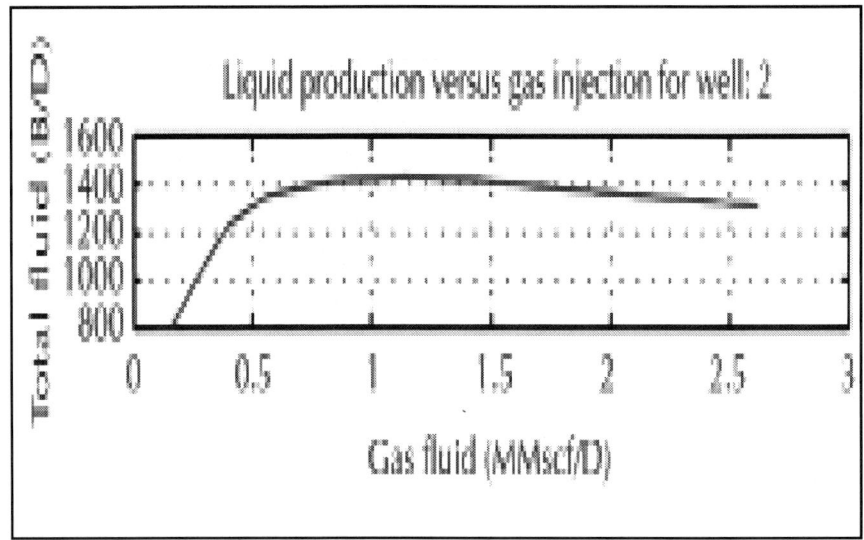

(a)

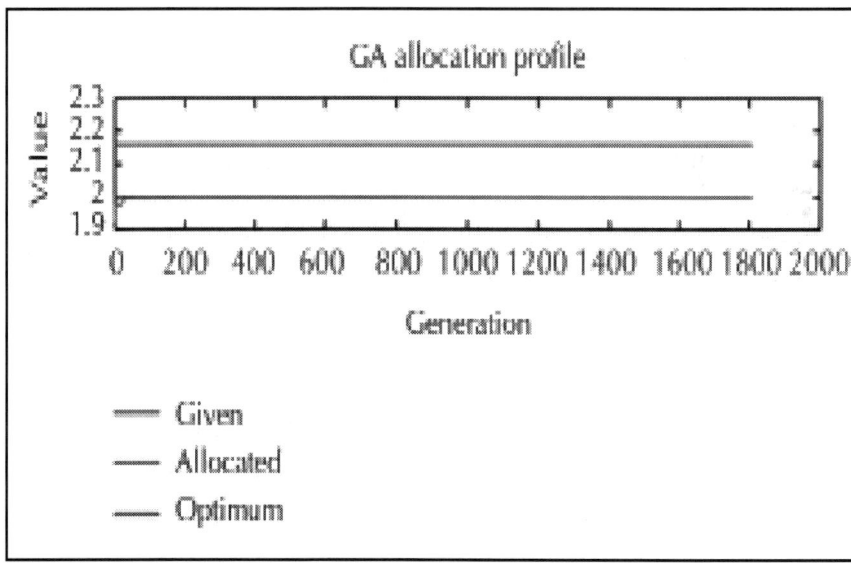

(b)

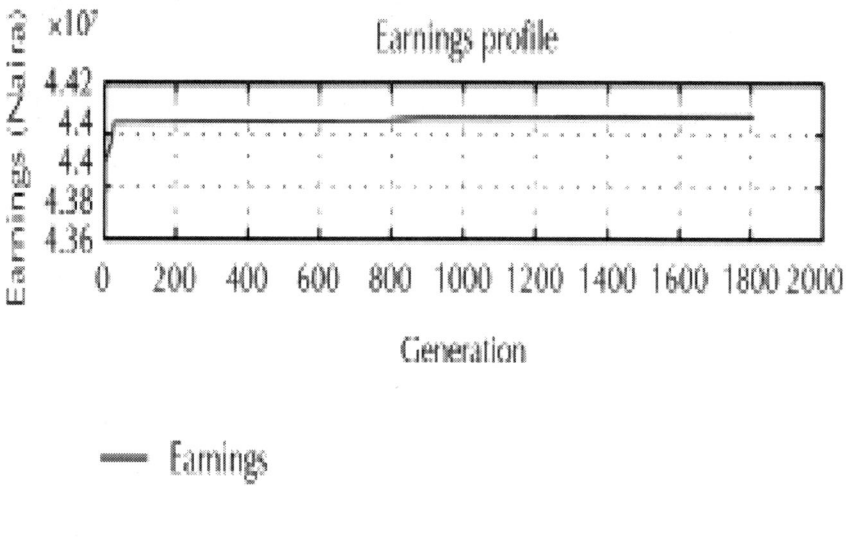

(c)

Figure 8: Well 2's allocation profile, wells 1-2 GA allocation profile, and wells 1-2 earnings (Naira) for a limited scenario.

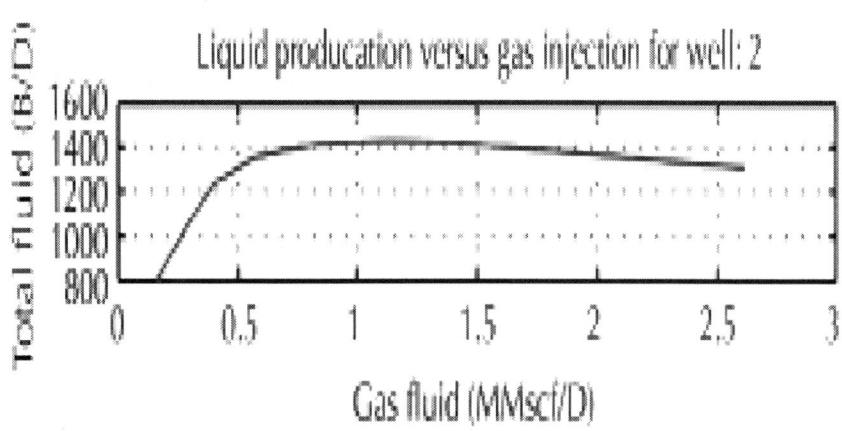

(a)

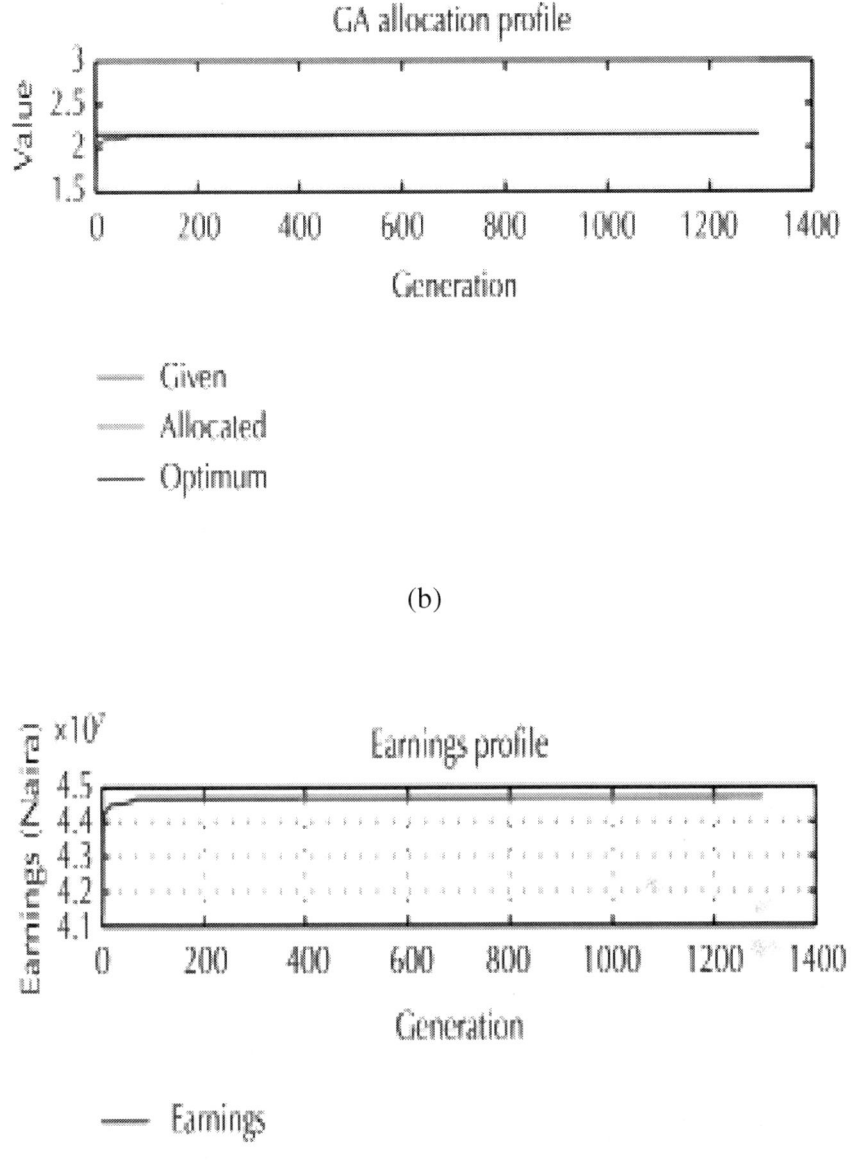

(b)

(c)

Figure 9: Well 2's allocation profile, wells 1-2 GA allocation profile, and wells 1-2 earnings (Naira) in an unlimited scenario.

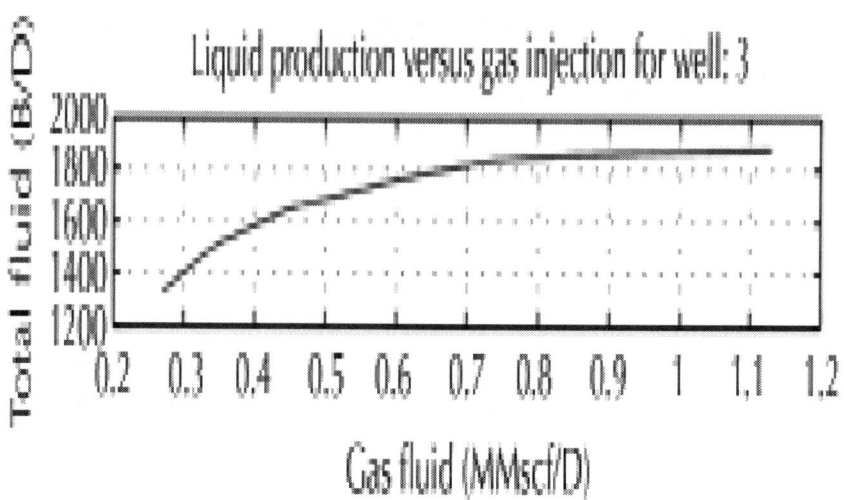

(a)

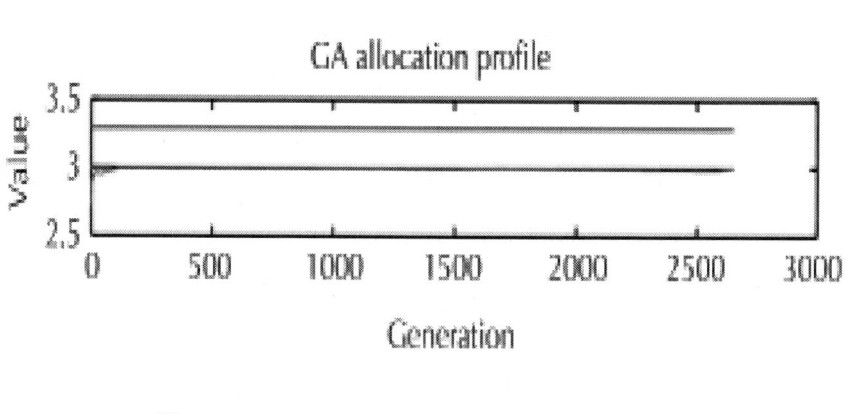

(b)

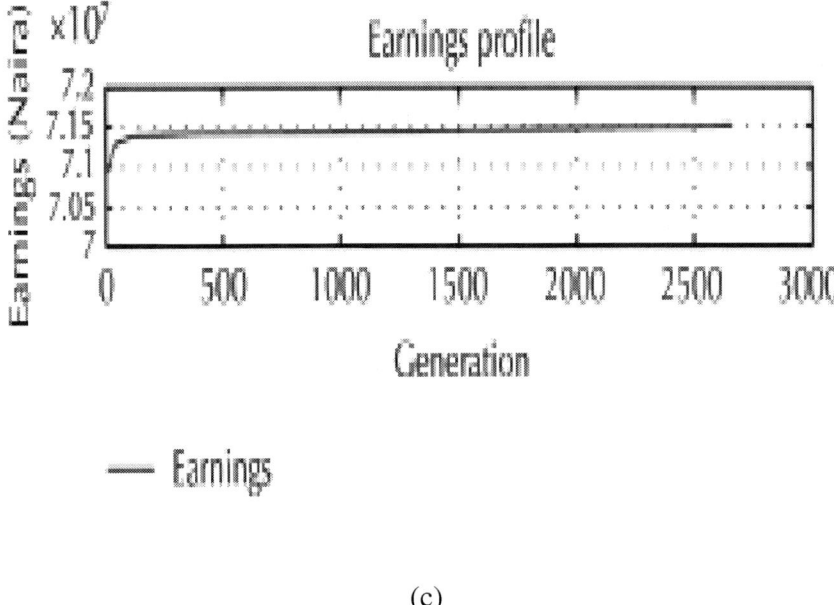

(c)

Figure 10: Well 3's allocation profile, wells 1–3 GA allocation profile, and earnings (Naira) in a limited scenario.

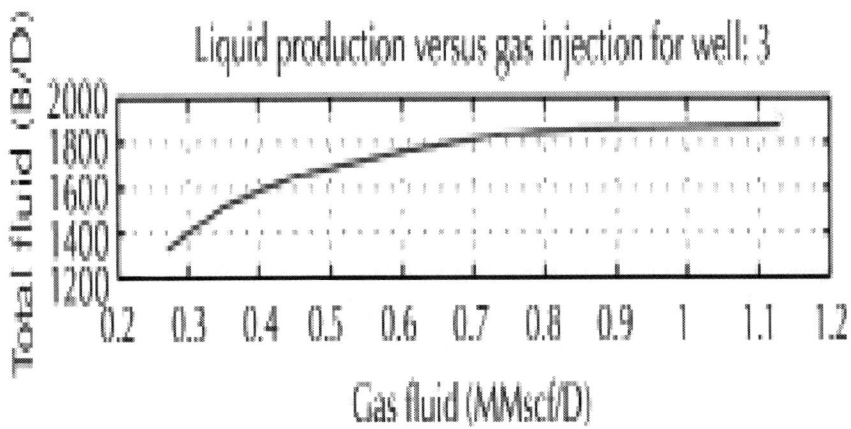

(a)

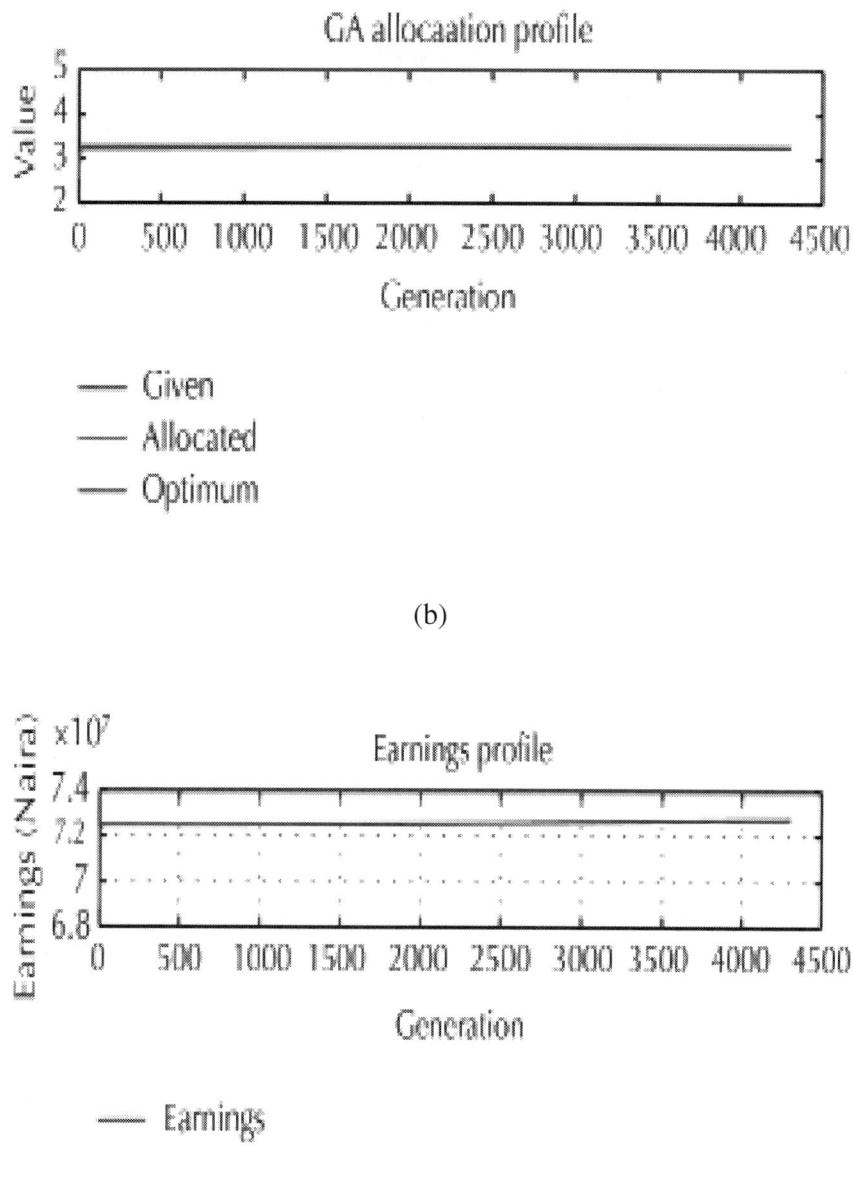

(b)

(c)

Figure 11: Well 3's allocation profile, wells 1–3 GA allocation profile, and earnings (Naira) in an unlimited scenario.

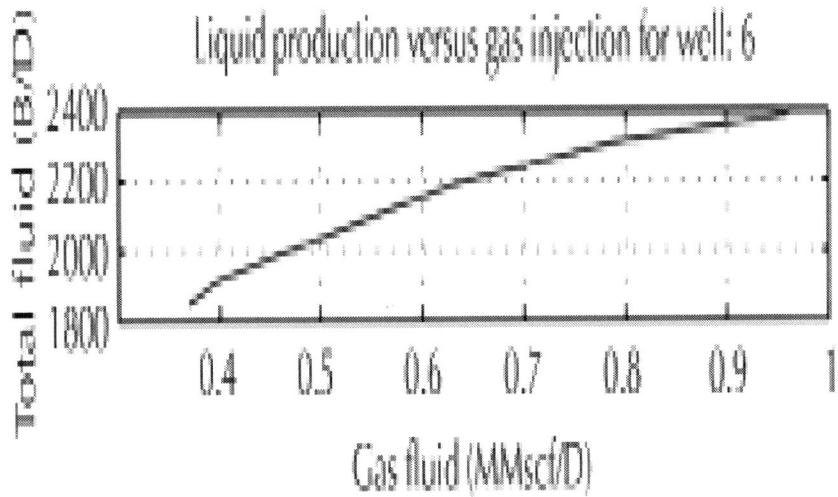

(a)

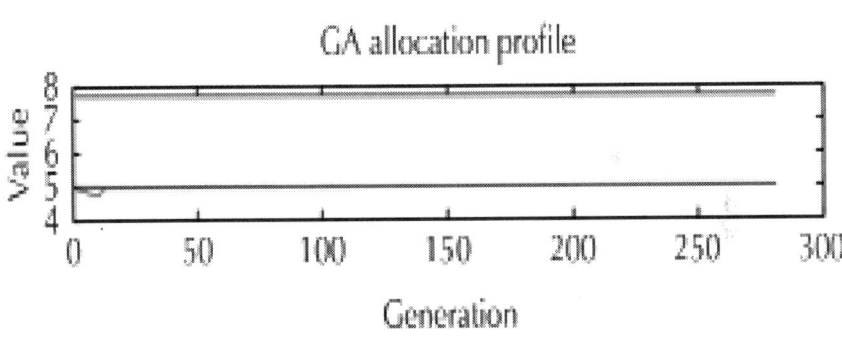

— Given
— Allocated
— Optimum

(b)

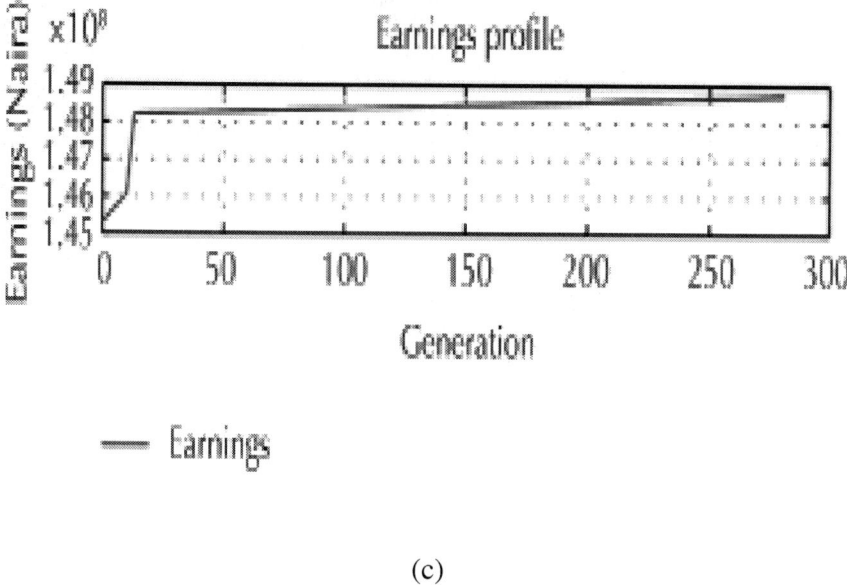

(c)

Figure 12: Well 6's allocation profile, wells 1–6 GA allocation profile, and earnings (Naira) for a limited scenario.

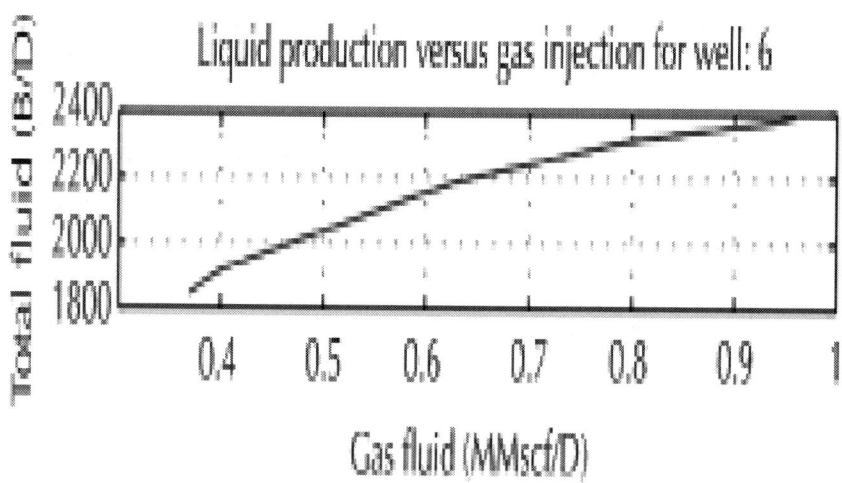

(a)

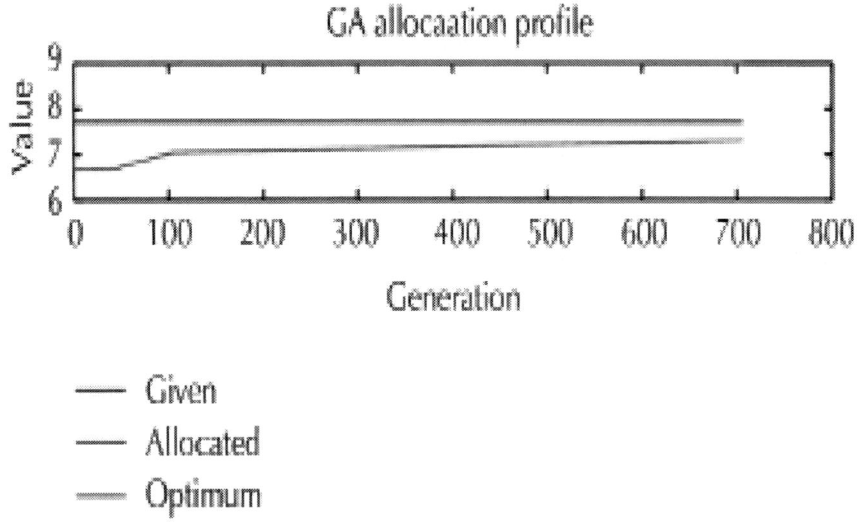

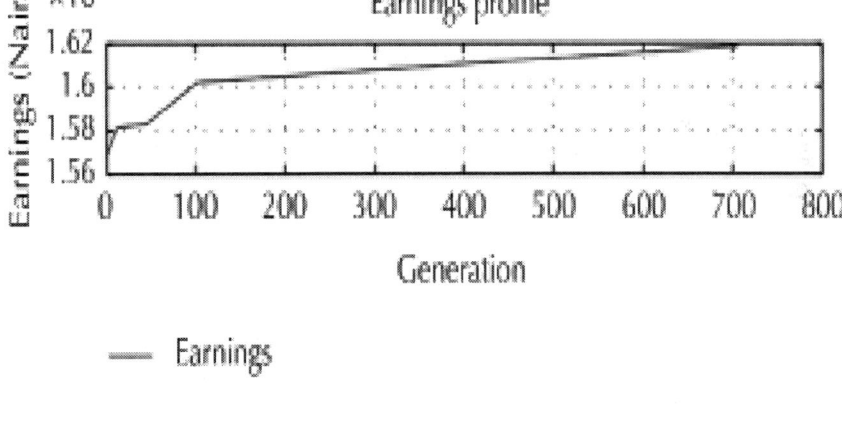

(b)

(c)

Figure 13: Well 6's allocation profile, wells 1–6 GA allocation profile, and earnings (Naira) for an unlimited scenario.

Results and Discussion

Based on the application of the proposed methodology in Section 4, results obtained for gas allocation to the six wells under the limited and unlimited scenario are presented mainly in graphical forms as given in Figures6–13. The behaviour of the proposed MANN-MIGA in effectively characterizing and allocating gas is discussed subsequently. Also discussed is the economic accrual obtained by methodology for different allocation configuration within defined generation run. Tables 2 and 3 show the best allocation values for the limited and unlimited scenario and also the average allocation.

Figure 6(a) displays the curve establishing the relationship between the allocated gas (in MMscf/D) and the oil produced (in B/D). It will be noted in this figure that the allocation of gases does not exceed the maximum point on the curve as already defined in given constraints otherwise it would have led to wastage since increasing allocation beyond the maximum point yields reduced oil production. The curve displayed in Figure 6(a) also echoes with the curves displayed in Figures 7(a), 8(a), 9(a), 10(a), 11(a), 12(a), and 13(a). The curves establishing the gas allocated versus oil produced for wells 1, 2, 3, and 6 are displayed in Figures 6(a) and 7(a), Figures 8(a)and 9(a), Figures 10(a) and 11(a), and Figures 11(a) and 12(a), respectively. It is observed that results illustrated in these graphs satisfy the expected constraints; that is, the allocation of gases does not exceed the value corresponding to the maximum oil that could be produced from that well.

In an attempt to allocate gas under the limited scenario in order to achieve maximum economic yield, Figures6(b), 8(b), 10(b), and 12(b) show how best gas was distributed among wells 1, 2, 3, and 6, respectively. The available gas in each of the considered scenarios was less than the sum of the gas value(s) corresponding to peak oil production for the number of wells under consideration. The MANN-MIGA results shown in Figures 6(b),8(b), 10(b), and 12(b) in allocating gas present the obtained best configuration that ensures maximum economic yield. The economic yield being referred to includes the combined accrual from sales of the respective oil produced and the unallocated gas using current market values.

Similarly, the combined MANN-MIGA was also used in allocating gas during an unlimited scenario (where the available gas for allocation exceeds the combined sum of the gas value(s) corresponding to peak oil production for the number of wells under consideration). Figures 7(b), 9(b), 11(b), and 13(b) show the allocation profiles for 1, 2, 3, and 6 wells during an unlimited scenario. Figures 7(b), 9(b), and 13(b) display the earnings based on the MANN-MIGA values and current market values for gas and oil.

In summary, four scenarios were considered for two conditions in each scenario, that is, limited and unlimited gas availability. The first scenario involved well 1 only, second scenario involved wells 1 and 2 only, and third scenario involved 3 wells, 1, 2, and 3 only, while fourth scenario involved the six wells. In each scenario, MANN-MIGA was tested for both limited and unlimited gas (in MMscf/D) availability. Figures 6(b), 8(b),10(b), and 12(b) show the allocation profile of MANN-MIGA in the limited scenario. As can be observed, gas quantity available for allocation, that is, given quantity (blue line), is less than the optimum value (red line) needed for maximum oil production. The allocation profile (green line) traces the given profile (blue line) in maximizing oil production. It should be noted that, in maximizing oil production and earnings (as shown in Figures 6, 8, 10, and 12), the criticality of the allocation configuration cannot be overemphasized. In essence, the configuration with the highest earnings is chosen.

Similarly, MANN-MIGA was also tested for the unlimited condition. As can be observed, Figures 7(b), 9(b),11(b), and 13(b) show the allocation profile under this condition. A further observation of the aforelisted figures shows that the given quantity (blue line) is greater than the optimum value (red line) needed for maximum oil production. Gas allocation (green line) is thus expected to follow the optimum value (red line) in generating maximum oil production values. The earning profiles for the different considered scenario under the unlimited condition are shown in Figures 7(c), 9(c), 11(c), and 13(c). In generating allocation configuration values and earnings, an oil price of N14, 000/B (Naira (N) is the currency for Nigeria where the work was carried out and based. The displayed value is in terms of Naira per barrel) and gas price of N400/MMscf were used.

The subsequent tables show the new values obtained in both the limited and unlimited scenarios. A critical observation of Figures 6(b), 8(b), 10(b), and 12(b) for the limited scenario and Figures 7(b), 9(b), 11(b), and13(b) for the unlimited scenario reveals some deviations.

As presented in Table 2, the MANN-MIGA under the limited scenario has a better allocation (about 99.94 on average) compared to the MANN-MIGA allocation values presented in Table 3 during the unlimited scenario (about 95.31 on average). As can be observed from Tables 2 and 3, three allocation terms are presented as further shown in the earlier considered figures. They are the optimum, given, and allocated values. The optimum value shows the maximum value that can be allocated to the well(s) under consideration guaranteeing the maximum oil that can be produced. The given value denotes the scenario under consideration, limited or unlimited, while the allocated value describes the ability of the MANN-MIGA in optimally distributing the given value for optimal economic/monetary yield. Our algorithm is therefore able to characterize the wells using a least square approximation method embedded into MANN and also allocate gas to the well combinations under consideration during limited (over 99% on average) and unlimited scenario (over 95% on average) using the MIGA as a stochastic optimization tool. The algorithm provided consistent results on multiple runs and at a fast run time.

CONCLUSION AND FUTURE WORK

This paper presents a characterization of oil wells and gas allocation in both the limited and unlimited scenarios such as presenting a cost-effective means for gas allocation for the oil and gas industry. While earlier works considered the characterization and allocation separately, this work has been able to both characterize and use generated values from the characterization in allocating gas. A mathematical model is obtained for characterizing the gas allocated and oil produced for the wells under consideration. Wells 1 and 2 were characterized using a quadratic equation while the rest were subsequently characterized using linear curve fitting techniques. A combined MANN-MIGA was adapted and applied extensively in evolving a relationship between the gas allocated and oil produced for the six wells considered. These values have been further used in computing the optimum gas allocation per well. The values generated show a remarkable allocation by our approach. The mild intrusive property of our GA arises from its ability to allocate the least possible gas value for maximum oil production. The approach neglects gas values that exceeded the maximum for optimal oil production. The MIGA was then used in generating optimized values which met given conditions and yielded improved economic returns. The MANN-MIGA thus proves useful to the oil and gas industry as it not only provides characterized equations, but also allocates gas based on preset conditions.

Future work might consider further fine-tuning of the algorithm for improved performance especially in optimizing gas allocation under the unlimited scenario. Also, other natured-inspired techniques like variants particle swarm optimization [17] and other recent stochastic algorithms [18] can be investigated for comparative analysis.

REFERENCES

1.   A. Codas and E. Camponogara, "Mixed-integer linear optimization for optimal lift-gas allocation with well-separator routing," European Journal of Operational Research, vol. 217, no. 1, pp. 222–231, 2012.

2.   T. Ray and R. Sarker, "Genetic algorithm for solving a gas lift optimization problem," Journal of Petroleum Science and Engineering, vol. 59, no. 1-2, pp. 84–96, 2007.

3.   E. Camponogara and P. H. R. Nakashima, "Solving a gas-lift optimization problem by dynamic programming," European Journal of Operational Research, vol. 174, no. 2, pp. 1220–1246, 2006.

4.   J. N. M. de Souza, J. L. de Medeiros, A. L. H. Costa, and G. C. Nunes, "Modeling, simulation and optimization of continuous gas lift systems for deepwater offshore petroleum production," Journal of Petroleum Science and Engineering, vol. 72, no. 3-4, pp. 277–289, 2010.

5.   M. Mahmudi and M. T. Sadeghi, "The optimization of continuous gas lift process using an integrated compositional model," Journal of Petroleum Science and Engineering, vol. 108, pp. 321–327, 2013.

6.   R. Sharma, K. Fjalestad, and B. Glemmestad, "Optimization of lift gas allocation in a gas lifted oil field as non-linear optimization problem," Modeling, Identification and Control, vol. 33, no. 1, pp. 13–25, 2012.

7.   O. G. Santos, S. N. Bordalo, and F. J. S. Alhanati, "Study of the dynamics, optimization and selection of intermittent gas-lift methods—a comprehensive model," Journal of Petroleum Science and Engineering, vol. 32, no. 2–4, pp. 231–248, 2001.

8.   E. P. Kanu, J. Mach, and K. E. Brown, "Economic approach to oil production and gas allocation in continuous gas lift," Journal of Petroleum Technology, vol. 33, no. 10, pp. 1887–1892, 1981.

9.   A. Ibitola and C. G. Monyei, "Genetic algorithm for students allocation to halls of residence using energy consumption as discriminant adaptive computing," Information Systems, Development Informatics & Allied Research Journal, vol. 3, no. 4, pp. 31–36.

10. C. G. Monyei and O. A. Fakolujo, "Optimized enhanced control system for the Unibadan's virtual power plant project using genetic algorithm," African Journal of Computing & ICT, vol. 6, no. 4, pp. 16–20, 2013.

11. D. Goldberg, Genetic Algorithms in Search, Optimization and Machine Learning, Addison-Wesley, Reading, Mass, USA, 1989.

12. A. O. Adewumi, B. A. Sawyerr, and M. M. Ali, "A heuristic solution to the university timetabling problem," Engineering Computations, vol. 26, no. 8, pp. 972–984, 2009.

13. A. R. Yildiz and F. Ozturk, "Hybrid enhanced genetic algorithm to select optimal machining parameters in turning operation," Proceedings of the Institution of Mechanical Engineers B: Journal of Engineering Manufacture, vol. 220, no. 12, pp. 2041–2053, 2006.

14. T. C. Nwaoha, Z. Yang, J. Wang, and S. Bonsall, "Application of genetic algorithm to risk-based maintenance operations of liquefied natural gas carrier systems," Proceedings of the Institution of Mechanical Engineers E: Journal of Process Mechanical Engineering, vol. 225, no. 1, pp. 40–52, 2011.

15. P. Guo, X. Wang, and Y. Han, "The enhanced genetic algorithms for the optimization design," in Proceedings of the 3rd International Conference on BioMedical Engineering and Informatics (BMEI '10), pp. 2990–2994, October 2010.

16. A. O. Adewumi and M. M. Ali, "A multi-level genetic algorithm for a multi-stage space allocation problem," Mathematical and

Computer Modelling, vol. 51, no. 1-2, pp. 109–126, 2010. View at Publisher ·View at Google Scholar · View at Zentralblatt MATH

17.   M. A. Arasomwan and A. O. Adewumi, "An adaptive velocity particle swarm optimization for high-dimensional function optimization," in Proceedings of the IEEE Congress on Evolutionary Computation (CEC ‹13), pp. 2352–2359, 2013.

18.   S. Chetty and A. O. Adewumi, "Three new stochastic local search algorithms for continuous optimization problems," Computational Optimization and Applications, vol. 56, no. 3, pp. 675–721, 2013.

Chapter 3

Decentralised Combined Heat and Power in the German Ruhr Valley; Assessment of Factors Blocking Uptake and Integration

Birte Viétor[1], Thomas Hoppe[2], and Joy Clancy[2]

[1]Department: Nature and Environment, Bonner Str. 100, Solingen, 42697, Germany

[2]Department of Governance and Technology for Sustainability (CSTM), Institute for Governance and Innovation Studies (IGS), University of Twente, Drienerlolaan 5, Enschede, 7522, NB, The Netherlands

ABSTRACT

Background

In Germany, the energy system is undergoing reorganisation from a centralised system based on fossil fuels and nuclear power to a sustainable system based on decentralised production and consumption of energy, the so-called *Energiewende*. Recently, there has been more attention to improving energy efficiency in those regions where conventional energy production activities and energy-intensive industries are located, such as the Ruhr area. Although the potential for decentralised combined heat and power (CHP) units is high in this region, local action plans show only modest developments for this technology. In this paper, we address this issue by answering the following research question: Which factors block the uptake and integration of decentralised CHP in the German Ruhr area's energy system?

Methods

The multilevel perspective (MLP) was used to analyse the state of system innovation in relation to the uptake and integration of decentralised CHP technology. Prior to the MLP analysis, a stakeholders' analysis was conducted to identify stakeholders' views, positions and experienced barriers regarding the uptake and integration of decentralised CHP technology. Data collection included review of text documents and conducting 11 interviews.

Results

Due to many regime barriers blocking niche development, the uptake of decentralised CHP technology is limited. Identified barriers relate to lack of market services and mismatches with user preferences, (sectoral) policies and industrial interests.

Conclusions

Observed barriers relate to (i) lack of market services such as financial means for making investments; (ii) user awareness such as unawareness and information deficit regarding the benefits of decentralised CHP to potential users, (iii) the presence of centralised district heating systems, (iv) policy issues such as lack of sufficient policies supporting diffusion of decentralised CHP units, legal stipulations from social housing policies that prevent housing cooperatives from becoming energy producers and district heating systems owned by public and private owners (via concessions contracts); (v) sector issues such as lack of skilled service-providing companies; and (vi) industrial interested such as the vested interests of the coal and gas industry. Moreover, many of the mentioned barriers seem interrelated, especially those concerning policy and finance available for making upfront investments.

BACKGROUND

Following the 2011 Fukushima nuclear disaster in Japan, the political decision was made in Germany to phase out nuclear power production by 2022 and to initiate a countrywide energy transition, the *Energiewende*. In short, it stands for a transition of the entire energy system in Germany which 'involves replacing, or supplementing, established technologies with new ones' [1] and thereby performing the 'inevitable shift away from cheap, centralized, largely fossil-based energy systems towards decentralised energy systems to a large extent based on renewable energy sources' [2] (translation by the authors). Next to increasing the share of renewable energy within the total primary energy supply to 50% by 2050, the program mentions the urgent need for action particularly in those regions where conventional energy production activities and energy-intensive industries are located. In these areas, there is large potential for using energy more efficiently. One way to foster energy efficiency is by supporting adoption of decentralised combined heat and power (CHP) units. The Federal Government of Germany strives to raise the share of combined heat and power generation to 25% by 2020 [3]. Since the opportunities for centralised district heating systems have already been utilised to a large extent, and are difficult to change over the short term for infrastructural and contractual reasons, raising the share of CHP would mean increasing the share of decentralised CHP units.

The Ruhr Valley in the Federal State of North Rhine-Westphalia (NRW) is a very striking example of a region for increasing energy efficiency levels since this is where many conventional energy production activities and energy-intensive industries are located. It is home to the majority of the coal power plants in Germany, and around one third of all German GHG emissions (313 million CO_2eq in 2010) are registered in this area. Furthermore, almost 30% of the national electricity demand is produced in this area (180 million kWh in 2011 in NRW; 608 million kWh in 2011 in Germany [4]), and around 25% of German final energy is used here. Moreover, 40% of the German energy demand for industrial processes originates in the Ruhr Valley [5].

The report 'Deutschlands Energiewende. Ein Gemeinschaftswerk für die Zukunft' ('Germany's energy transition. Collaborative work for the future'; translation by the authors) that was published by the Ethical Commission on behalf of the Federal Government in May 2011 underlines the important role decentralised governments have in implementing *Energiewende* policies. For instance, the report points out that it is the municipalities, rather than the central or state government, that have the planning authority when it comes to locating sustainable energy production utilities. Furthermore, municipalities are responsible for public buildings and public transport; via their municipal energy suppliers and housing societies, they are delivering electricity and heat [6],[7], and potentially, they can have policies in place to influence energy consumption by local stakeholders. Moreover, it is at the local level that environmental matters (including energy transition) manifest and citizens engage with government [8]. Hence, cities and regions can become powerful promoters of sustainable transitions [9]. Against this background, it is reasonable to discuss the energy transition approaches from a situational perspective as we deem the situational context and hence specific regional conditions, of great importance to strategies to effectuate energy transitions in cities and regions. In particular, situational conditions might favour adoption and rollout of one particular energy transition pathway, whereas it might disfavour alternatives pathways.

It is for these reasons that we explore the potential uptake and integration of decentralised CHP in the Ruhr Valley. Although the *Energiewende* is widely known to support diffusion of wind and solar systems for power production; in particular, it focuses on electricity generation from renewable sources; it also highlights the need for action towards using energy more efficiently in those regions where conventional energy production activities and energy-intensive industries are located, notably the Ruhr Valley. Given its very nature, the industrial, highly populated urban region of the Ruhr Valley offers high potential for further take up, integration and upscaling of decentralised CHP. In NRW, the share of CHP total electricity production, however, amounted to only 10% [10].[a] Hence, it is far removed from the national target of 25%.

[a][19] EU Tech Energie & Management (2008) NRW-Klima2020 - Beitrag Nordrhein-Westfalens zur Erreichung des nationalen Klimaschutzziels. Aachen.

Analysis of the local climate change protection action plans addressing the uptake and integration of decentralised CHP units provides an overview on the importance that is given to the uptake of this technology by nine Ruhr Valley cities (see Table 1).

Table 1: Overview is policy action plans Ruhr city governments have prepared regarding CHP

City	Year of origin of policy action plan	Inhabitants	Actions regarding decentralised CHP
Bochum	2009	373,000	Feasibility study: integration of small CHP (buildings) in the regional supply concept ('virtual power plant').

Bottrop	2011	116,000	Local heating grid Kirchhellen: construction of a CHP plant of 15,000 kW with local heating grid on the basis of renewable energies (biogas, wood).
Dortmund	2011	580,000	Potential study of utilising CHP units in private households and the commercial sector.
			Block-heating station-'Push' in municipal properties.
Duisburg	2009	488,000	None.
Essen	2009	573,000	Integration of urban development contracts into town plans to regulate use of CHP.
Gelsenkirchen	2011	256,000	Block-heating station-'Push' in municipal properties.
			Pilot project: 'Extension of CHP'.
			Local heating grid in interconnection (potential analysis for block heating stations).
			Combination of geothermal energy and micro-CHP.

Oberhausen	2012	212,000	Heat supply in the form of heat contracting for private house owners.
			Decentralised CHP for the local heating grid in Barmingholten.
Recklinghausen	2012	118,000	Extension of (renewable energy) CHP in the city area.
Witten	2013	98,000	'Push' for combined heat and power.

Viétor *et al.*

Viétor *et al. Energy, Sustainability and Society* 2015 5:5, doi:10.1186/s13705-015-0033-0

Eight out of the nine cities have formulated measures that aim at increasing the share of heat and power produced in decentralised CHP units.[b] The actions planned by these eight municipalities differ in various ways. Whereas some municipalities mention in their local action plans that decentralised CHP should be used in municipal buildings, others target local stakeholders, like households and commercial firms. Different policy actions are mentioned such as constructing apartment block-heating installations, integration of decentralised CHP units in urban development planning and establishing contracts with private home owners for the provision of heat. In some municipalities, decentralised CHP is seen as a means to feed in decentralised-produced heat into the existing district heating grids. In this way, decentralised CHP can be seen as a means to 'extend' existing (district) heating grids. Not surprisingly, multiple Ruhr Valley local authorities deploy 'extension of CHP' actions in their respective programs due to the presence of centralised district heating grids. Studies regarding the technical and economic potential or feasibility of decentralised CHP have been prepared in four municipalities, and plans for constructing decentralised CHP units are mentioned by at least four (other) municipalities. The ways the nine city governments address the uptake and integration of decentralised CHP shows little coherence: between the nine municipal plans, there is little alignment in the formulation of goals and policy actions. In summary, the policy actions to spur the uptake and integration of decentralised CHP by the nine Ruhr Valley municipalities give the impression that uptake of decentralised CHP is at a rather early stage.

[b]There is one exception: the city of Duisburg. The reason is that the district heating grid already supplies heat to the inhabitants of the municipality. The city already goes beyond the national objective to raise the share of electricity produced in CHP plants to 25%.

These findings stimulated us to find out why the uptake and integration of decentralised CHP is currently not at a more developed stage. *A priori*, we would expect that there would be a large-scale uptake and integration of CHP in the Ruhr Valley, but we do not see it in reality. Therefore, we aim to identify the factors that might explain why the potential for decentralised CHP is not being realised. The main research question of this paper is: *Which factors block the uptake and integration of decentralised CHP in the German Ruhr area's energy system?*

In order to answer the question, we use stakeholders' and systemic analytical perspectives for sustainable transitions. In line with Truffer and Coenen [9], we view sustainability transitions as political projects where stakeholder interrelationships play a crucial role in the transformation process. Therefore, we conduct a stakeholder analysis to identify relevant stakeholders' roles, positions, interrelationships, views and experienced barriers concerning the uptake and integration of decentralised CHP. Next, we analyse the systemic status quo regarding the uptake and integration of decentralised CHP into the energy system of the Ruhr Valley by using the multilevel perspective (MLP) framework for transitional change [11],[12].

Background: the Ruhr Valley region context

The Ruhr Valley is one of the biggest urban agglomerations in Europe with 5.2 million inhabitants. The region can be divided into 11 urban districts (Kreisfreie Städte; translation by authors) (Duisburg, Mülheim a. d. Ruhr, Oberhausen, Essen, Bottrop, Gelsenkirchen, Herne, Bochum, Dortmund, Hagen und Hamm) and four counties (Kreise) (Wesel, Unna, Recklinghausen und Ennepe-Ruhr). These in turn are composed of 43 independent cities and municipalities within the county. Figure 1 presents the siting of the region in Germany and the siting of administrative divisions within the Ruhr Valley [13].

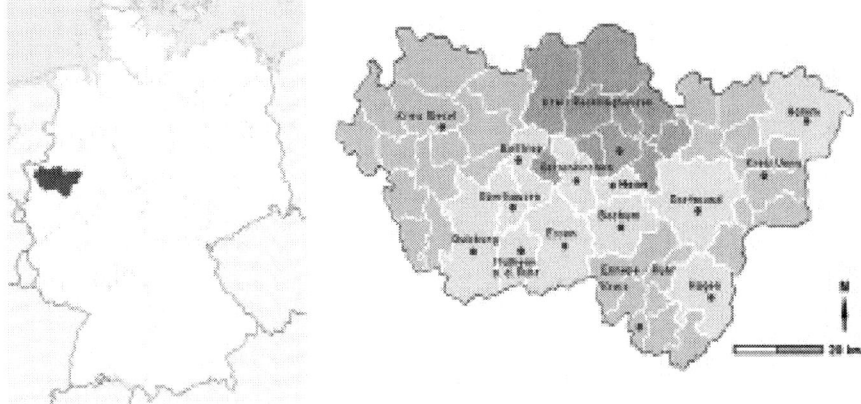

Figure 1: Siting of the Ruhr Valley region in Germany, and the Ruhr Valley's administrative divisions.

The coal deposit of the Ruhr area provided the natural precondition for the development of the biggest coal and steel industry in Europe. Thus, historically, the Ruhr area was the hotspot for steel and chemistry industries - economic sectors that are strongly linked to greenhouse gas emissions. During 150 years of industrialization, the region underwent strong spatial changes as settlements expanded and economic activity accelerated. Industrial and socio-economic institutions were essentially designed to serve the mining industry. While in the second half of the 19th century 280 coal mines were active in the region, this number diminished dramatically. In 2009, only four mines were left [14]. Nowadays, more than 70% of the population in the area is employed in the tertiary sector. This impressive development from an industrial to a knowledge society could not solve all problems that are linked to the 1950s crisis in the coal industry (e.g. high unemployment, financial debts) [15]. Some cities in the Ruhr area are nowadays among those in Germany with the highest unemployment rates (e.g. Duisburg, Dortmund). The unemployment rate contributes to the weak financial situation of the municipalities[c] which limits the scope for action.

[c]For example, in 2012, the municipality of Oberhausen was the most indebted municipality in Germany [15].

Background: decentralised CHP technology

Low efficiency during the electricity generation processes results in energy being lost in the form of waste heat [16]. Collecting waste heat for reuse combined with electricity generation is a way to use primary energy more efficiently [17]. It is the combination of the electrical generation and heat production process that forms the basis of the CHP concept [18]. The overall efficiency can reach up to 90%, whereas single-electricity generation only reaches efficiency levels of 30% to 60% [19]. Waste heat from the exhaust gases, or used steam, is recovered for instance to produce hot water for the use in district heating schemes (in large-scale CHP systems) or to heat spaces in buildings (using smaller decentralised CHP plants; [16]). Three categories for use of CHP have been defined by the German Federal Ministry of the Environment, Nature Protection and Nuclear Safety, one of them being decentralised CHP. The latter is typically used to supply heat and power for consumption in residential or commercial buildings [19]. In this paper, 'decentralised CHP' relates to CHP units used at the district level, street level or at the level of (multistorey) housing complexes.[d]

[d]We would like to stress, however, that in the context of the Ruhr area there is no common conceptualization of 'decentralised CHP'. From our interviews with local practitioners, we conclude that there seem to be multiple understandings of 'decentralised CHP' varying among (types of) stakeholders and contexts.

Decentralised CHP offers considerable benefits as compared to conventional ways of heat and power production. First, it offers improved energy efficiency and preserves non-renewable energy reserves. Second, due to more efficient conversion of primary energy, reductions in ground-level particle and gas emissions and pollution can be realised, hence improving urban air quality [16]. Third, both fossil and renewable sources can be used in decentralised CHP plants; hence, technology can make an important contribution to reducing greenhouse gas emissions and the primary fossil fuel energy consumption in the future energy system [19].

Background: decentralised CHP technology uptake in the Ruhr Valley energy system

The integration of decentralised CHP into the German energy system forms an important cornerstone of the energy policy of the Federal Government of Germany. The Integrated Energy and Climate Protection Program (Integriertes Energie- und Klimaschutzprogramm) (2007) of the Federal Government of Germany includes the political target to raise the share of combined heat and power generation in Germany from the current level of about 13% to 25% by 2020 [3]. The important role CHP plays in the framework of the energy system transition was affirmed in a renewal of the respective law (Act on CHP) and the creation of different funding policies (e.g. the Act on Renewable Energies). Since 2003, power and heat generation via CHP has increased by 20% and 8%, respectively. In 2005, 18.2 TWh power was produced by CHP in NRW. Related to the entire power production in NRW (ca. 180 TWh/year), the share of CHP amounted to only 10% [10].

In January 2013, the Federal State NRW adopted the first Climate Protection Act in Germany which contains legally binding climate protection objectives. The act aims at contributing to the achievement of the national climate change mitigation goals and acknowledges the potential that decentralised CHP has in helping reach these objectives. It further acknowledges the significant responsibility the region has in this respect, notably when compared to other German States. According to the Climate Protection Act, by 2020, GHG emissions are to be reduced by 25% relative to the 1990 level. By 2050, a reduction of 80% relative to the 1990 level is targeted. NRW also adopted the national CHP objective based on which the state government developed a program, the 'CHP Impulse Program', to increase the share of decentralised CHP in the area. The CHP Impulse Program, together with the national political conditions, makes the current political framework conditions seem supportive towards the integration of decentralised CHP into the energy system in NRW.

Recent studies on the potential of CHP in NRW (May 2011) and on district heating in the Ruhr area (May 2013) have stressed the significant potential for this technology in meeting policy goals. Both studies argue that it is economically viable to extend the district heating system, in particular to supply heat to more residential areas in the Ruhr Valley. However, the large-scale, centralised CHP plants that supply heat to the district heating grid are emphasised. At the same time, the studies mention some barriers that could have strong impacts on the operating efficiency of the district heating system. Namely, due to improvements in energy efficiency and insulation of buildings, it is predicted that there will be a decrease in demand for heat and electricity. Furthermore, demographic changes will have an impact on the number of customers demanding heat and electricity. At the same time, many coal power plants are expected to close down in the near future, because the profit margins are decreasing rapidly. Consequently, less heat will be generated and supplied to the grid. Interestingly, these barriers do not seem to have a significant impact on the formulation of recommendations for the future energy system.

The study 'Investigation of the potential of CHP in Northrhine-Westphalia' (in German:*Potenzialerhebung von Kraft-Wärme-Kopplung in Nordrhein-Westfalen*; translation by the authors)[20] states that an increase in the number of decentralised CHP plants will lead to decreased connection rates to the district heating grid which in turn will have a negative impact on the cost effectiveness of the latter. Although this sounds logical, it did not lead to any critical reflection by policy makers on how to overcome this potential conflict between centralised and decentralised supply of energy. The study 'Perspectives of district heating in the Ruhr Area until 2050' (in German: *Perspektiven der Fernwärme im Ruhrgebiet bis 2050*; translation by the authors) [21] mentions briefly that there is a need for more flexibility and more heat sources to be able to maintain the economic viability of the district heating network. Responding to this point is very important as decentralised CHP units could theoretically ensure this flexibility and at the same time provide more heat sources.

In summary, there are five reasons why the potential for large-scale adoption of decentralised CHP in the Ruhr Valley should be considered worthwhile exploring. First, the two studies mentioned above underline the high potential regarding the further integration of CHP into the energy system as well as connecting to the existing district heating grid [21],[22]. Second, the high-geographical proximity of cities in the Ruhr Valley combined with the high-population density in the area offer good preconditions for the efficient use of CHP - both centralised and decentralised. Third, the decentralisation of energy production and consumption is a stated political objective (in the*Energiewende* program) and as such could provide a window of opportunity for policymaking to support decentralised CHP uptake and integration in the future energy system. Moreover, promoting energy efficiency is viewed as an urgent priority *vis-à-vis* the future of energy transition in Germany. Although supply of heat and power from cogeneration and energy efficiency in buildings have taken a backseat in the German energy transition agenda, these sectors have recently been identified by researchers as having significant potential [22]. Fourth, increased uptake of decentralised CHP allows for greater flexibility in responding to fluctuations in energy demand and supply. Fifth, decentralised CHP also allows for local use of renewable energy sources. For this reason, CHP can play an important role in the envisaged energy system transition in Germany and contribute to attaining climate change policy goals. In conclusion, decentralised CHP could potentially offer substantial benefits to energy producers, end users, policymakers and governmental actors.

Theoretical framework: using the multilevel perspective to analyse regional energy transitions

The challenge for the uptake and integration can be viewed as fostering a socio-technical and sustainable transition. A *socio-technical transition* is a set of processes that leads to a fundamental shift in socio-technical systems (e.g. [23],[24]). A socio-technical system (such as the energy system) consists of '(networks of) actors (individuals, firms, and other organisations, collective actors) and institutions (societal and technical norms, regulations, standards of good practice), as well as material artefacts and knowledge. The different elements of the system interact, and together they provide specific services for society' [25]. The systems concept highlights the fact that a broad variety of elements is tightly interrelated and depends on each other [26].

Sustainability transitions, for instance those intended to attain the status of a low-carbon society, are also considered by researchers to be socio-technical transitions. Energy transitions, for example the *Energiewende*, can also be viewed as *sustainability transitions*. These are long-term, multidimensional and fundamental transformation processes through which established socio-technical systems shift to more sustainable modes of production and consumption. A key feature of sustainability transitions is that guidance and governance often play a significant role in the transition [27], when decisive interventions from state and non-state actors are needed to overcome the inertia and lock-in typically found in the prevailing socio-technical systems [26],[28],[29].

Multilevel perspective on system innovation and transitional change

Currently, the theoretical framework most used to understand and systematically analyse socio-technical transitions, in particular the energy transition (e.g. [30]), is the MLP [11],[12]. The framework has been developed based on findings from evolutionary theory and systems analysis. According to Geels 'the stability of established socio-technical configurations results from the linkages between heterogeneous elements' [12], p. 1259. For MLP, these are linkages between three conceptual levels: macro, meso and micro.

On the *macro level* 'landscape events' occur. Landscape is associated with the material context of society. It is made up of various macro factors such as oil prices, interstate geopolitical relationships and forthcoming events (like treaties and wars), political and governmental coalitions, cultural values and major environmental problems. The socio-technical landscape forms the external context for action of, and interaction between, actors.

The *meso level* is referred to as the 'socio-technical regime'. These regimes encompass social and institutional rules that enable and constrain activities between actors [12]. These rules are related to several institutionalised factors, such as markets, user preferences, (sectoral) policies, industries, science, culture and technology. As a rule of thumb, socio-technical regimes only change incrementally and contain defence mechanisms to fend off attempts to replace them with alternatives, typically radical innovations developed at the micro level in 'socio-technical niches'.

The *micro level* is referred to as the analytical level in which 'socio-technical niches' develop. The niche is one of the central concepts in transition research. Niches form protective spaces in which radical innovation can develop, while being protected from regime defence mechanisms. An important question regarding the role of the niche is upscaling (e.g. addressing the question how to increase the take up and integration of the niche within society). Smith and Raven [31] differentiate three functions for niche protection: shielding, nurturing and empowering. Shielding refers to processes that hold at bay selection procedures from mainstream selection environments. Nurturing refers to processes such as learning, networking and expectation formation. Finally, empowering refers to the process that makes niche innovations competitive within unchained selection environments ('fit-and-conform') and processes that restructure mainstream selection environments in ways favourable to the niche ('stretch-and-transform'). Political, administrative, managerial and academic interest in how to encourage (sustainable) transitional change have led to the development of managerial transitional change frameworks, notably the Strategic Niche Management (SNM) [32],[33] and Transition Management (e.g. [34],[35]).

The interlinked character of the macro, meso and micro levels means that regimes are embedded within landscapes and niches within regimes. Innovations (and hence attempts to bring about transitional change) take off in niches in the context of existing regimes with their specific problems, rules and capabilities. Thus, in a transition process, interactions between dynamics exist between the three levels.

Theoretically, the systemic dynamics that result in transitional change follow a typical pattern. Landscape events (like the 2011 Fukushima nuclear disaster) create pressure on socio-technical regimes which result in problems that regimes cannot solve from within. Solving these problems via incremental regime optimization will not suffice to solve these problems and creates opportunities for alternative radical innovations with the potential to overthrow the current regime. This provides opportunities for new innovations that develop in niches and are supported by social networks (often including 'regime outsiders') that are active in niche formation activities. After iterative sets of niche experiments (e.g. by organising demonstration projects), innovations mature and have the potential to gain a foothold in the existing socio-technical regime. If successful, it can eventually replace the existing regime and hence the socio-technical system as such [36]. When replacement of an existing regime by a novel regime takes place, one can speak of *system innovation*. When this concerns a radical innovation, one can speak of *transitional change*. Breakthroughs of radical innovations consequently depend on interactive systemic processes within and between the macro, meso and micro levels (i.e. between landscape events, the socio-technical regime and niches). In general, transitions are context dependent [12].

Although often used in academic studies of transitions, MLP and the related analytical frameworks that suggest 'management of transitions' (e.g. SNM, TM) have been criticised for failing to address context-related (regional and local) factors that influence the way in which innovations develop and transitions manifest (see e.g. [26],[37],[38]), notably addressing agency [39]-[41], politics and power [42]-[44]; the role of user groups in consumptive action [37],[45]; the role of civil society in transitions [46],[47]; policy [48],[49]; and geographic-bound factors such as comparative institutional, socio-economic or infrastructural advantages that differ substantially from one geographic administrative jurisdiction to another [50]. When analysing energy transitions at a regional scale, identifying these factors is important because one needs to first understand the context, culture and historical pathways that explain the establishments of key institutions, interstakeholder power relations and comparative institutional (dis-) advantages. Based on this perception, the next step is then to understand the functioning of the socio-technical systems, before one attempts to plan and 'manage' transitional change. Based on this criticism, Geels recently [51] furthered the conceptualization of energy transitions by incorporating agency, power and politics. In relation to attaining low-carbon societies, he [52] states that there should be more academic attention to resistance by incumbent regime actors to fundamental change than continuing to overemphasise the potential of a great many 'green' innovations. Regarding future research agendas, this should be understood as shifting away from studying niche development to studying regime power and politics as defensive mechanisms to prevent radical change from happening. Hence, the unit of analysis should be the regime, not the niche. In this paper, we adopt Geels' view, although our initial focus is on a particular niche (decentralised CHP). However, the main part of our analysis addresses regime forces (see the 'Methods' section).[e]

[e]For this reason, we decided to include conducting a stakeholders' analysis in our research design. Stakeholder analysis reveals much of existing regime barriers, agency relations, power and politics.

Furthermore, the paper contributes to the academic literature on the energy transition (strategies towards attaining low-carbon goals) in cities (e.g. [53]-[55]; and the thematic issue of which this paper is part [56]).

METHODS

This paper presents the case study of the Ruhr Valley, which will be analysed in two ways, in which the results of the first analysis will be used as the input for the second analysis. First, a stakeholders' analysis will be conducted to identify key stakeholders, their positions in the regional energy system and barriers they experience in relation to the uptake and integration of decentralised CHP units. Second, a systemic analysis will be conducted using the MLP framework. Results from the stakeholders' analysis (notably identification of regime barriers) will be used as input for the analysis with MLP.

Case Selection

The Ruhr Valley was selected for its urban and industrial regional character. Theoretically speaking, these characteristics would be expected to benefit the take up of energy efficiency technological pathways, in particular decentralised CHP. This makes our case different to other areas in Germany which have adopted renewable energy technologies on a large scale (in particular wind and solar energy) because they are more rural and hence spatially more suited to spur production of green electricity. Thus, case selection follows the criterion of the regional socio-economic and spatial character. Selecting the Ruhr Valley as a regional case study should be seen as selecting an extreme case[57] for Germany as no other region compares readily to the Ruhr area. So, hypothetically, the highly industrial and urban character of the region, with potential for use of residual heat and high local demand for heat and power, should favour uptake and integration of decentralised CHP. As this study presents a (single) case study, external validity (and hence representativeness) is low. Nonetheless, potentially regions that also feature the previously mentioned characteristics can readily be compared with the Ruhr area. However, this is not the objective of this paper which only studies the one region.

ᶠHere we assume a relationship between the spatial type of region and the uptake of CHP. For classifying a region, we assume an ordinal scale which on the extreme can be classified as 'rural/non-industrial' and on the other side as 'urban/high industrial'. Hence, the Ruhr Valley region can be discerned as an 'extreme case' [58] since it is characterised by its very urban and highly industrial nature.

Stakeholders' Analysis

A stakeholder analysis is conducted to generate knowledge about the stakeholders involved with decentralised CHP 'so as to understand their behaviour, intentions, interrelations and interests; and for assessing the influence and resources they bring to bear on decision-making or the implementation process' [58], p. 338. The stakeholder analysis consisted of the identification of stakeholder characteristics, views and experienced barriers *vis-à-vis* the uptake and integration of decentralised CHP. Data collection concerned 11 qualitative - that is face to face, in-depth - interviewsᵍ with representatives from different stakeholder groups relevant to the uptake of decentralised CHP. A semi-structured questionnaire was used during the interviews. Questions raised addressed the status quo of decentralised CHP uptake in the Ruhr Valley, the future role of this technology in the regional energy system, interaction with district heating, experienced barriers, niche development, policy support strategies and (other) stakeholders' influence on development of this technology in the Ruhr Valley region. Addressing these issues, in particular, the barriers that hinder the uptake and integration of decentralised CHP provide key information necessary for understanding the forms of resistance exercised by the incumbent socio-technical energy regime [1].

ᵍThe interviewees were directors of environmental and climate protection units in city governments, energy consultants at the regional and local level (including a representative from the National CHP Association), managers of energy suppliers, a representative of a public utility company, representatives of an energy agency and a research institute, and a CHP project manager from city government.

Systemic Analysis of the Status Quo Using the MLP Framework

In this paper, decentralised CHP is treated as a 'niche' (or technology with the potential to become a radical innovation) using the MLP conceptualization, although we argue that this technology has the potential to become a radical innovation and stimulate the (sustainable) transition in the energy system. In order to analyse niche development landscape events, regime factors and niche development will be identified. In common with MLP, the focus of our analysis will be *identification*of regime forces that tend to block decentralised CHP niche development.

Data used for conducting the analysis included the results from the stakeholder analysis (to a large extent this had identified the barriers) and from an analysis of documents. Next, the data was categorised into three clusters representing the three MLP levels of landscape (macro level), regime (meso level) and niche (micro level). After which, the interlinkages between the three levels were identified.

MLP is used a lot to analyse historical cases [26]. However, in this paper, we use MLP as an analytical tool to identify factors (and their interlinkages) that hinder or support development of a contemporary niche. Hence, MLP is used to assess the status quo of regime forces that influence the niche development of decentralised CHP.

RESULTS

The results of the Ruhr Valley case study are presented in two parts. First, the results of the stakeholder analysis are presented, in terms of stakeholders' views, positions, interrelations and experienced barriers *vis-à-vis* the uptake of decentralised CHP. After which, the results of the systemic analysis using MLP are presented.

Results of Stakeholders' Analysis

Table 2 presents an overview of the stakeholder characteristics and views regarding the uptake of decentralised CHP. In the table, stakeholders, their positions, interrelations with other stakeholders, opinions towards the uptake of this technology and experienced barriers are presented. Table 2 reveals that consumers - be they private households, the commercial sector or housing associations - are sceptical towards implementation. In the next section, barriers experienced by stakeholders will be discussed in more detail.

Table 2: Stakeholders characteristics and views regarding uptake of decentralised CHP

Stakeholder	Position and function	Interrelation with other stakeholders	Opinion towards uptake of decentralised CHP	Experienced barriers
Energy supply companies (e.g. RWE/E. ON)	Produce and supply energy and operate some of the local heating grids via long-term concessions.	Supplies heat and power to end consumer via DSO. Heavily regulated by government.	Negative	Combining sufficient regional and local heat demand, failing centralised heating plants (momentarily absence of heat supply).
Local governments/ cities	Run local action plans to support uptake of decentralised CHP. Own public utility companies that in turn operate local heating grids.	Serves the common interest. Owns public utility companies. Relations to other stakeholders via local decentralised CHP action plan.	Positive/ neutral	Financial deficits, lack of information, lack of central government (climate) policy to support decentralised CHP, uncertainty related to elections, and lack of policy focus.

Providers of decentralised CHP	Produce and sell decentralised CHP units and systems.	Provide CHP unit to end users, housing corporation or private contractor to exploit the CHP unit.	Positive	Existing district heating grid, lack of information, lack of central government (climate) policy goals to support decentralised CHP, strong coal and gas lobby in NRW policymaking venues, elections and lack of policy focus.
CHP unit system installers	Provide installation and maintenance services that are used by end consumers.	Installs and maintains CHP units as commissioned by end users, housing corporation or private contractor to exploit the CHP unit.	Neutral, albeit hardly aware of benefits and potential	Status of a historical monument of buildings and its legal implications, lack of information, elections and lack of policy focus, lack of market (support) services, due to small size little innovation receptive capacity.
Private contracting parties	Operate commercial decentralised CHP units and make contracts with end users.	Contract with end users or housing corporation.	Positive	Lack of information, many barriers from other stakeholders indirectly harm contracting parties.

Consumers (e.g. households)	Use of energy produced by decentralised CHP units (heat and/or electricity).	Contractually related with CHP unit installer, contracting party, (in some cases) housing corporations, consultants, and the DSO. Might be connected to local government via decentralised CHP action plan (e.g. recipient of subsidy).	Sceptical, albeit hardly aware	High upfront investments, lack of financial governments support, monumental status of buildings, existing district heating grid, lack of information, habits of using energy equipment, transaction costs that go with CHP unit registration and using subsidies, elections and lack of policy focus.
Social housing corporations	Adopt, own or rent and operate decentralised CHP units from which the produced energy is to be used by tenants of the housing corporation.	Contractually related with CHP unit provider, CHP unit installer, contracting party, DSO and local government, energy company.	Positive	Lack of information, transaction costs that go with CHP unit registration and using subsidies, social housing corporations legal entitlements and limitations.
Consultancy/ advisory agencies	Provide advice clients seeking information on adopting, using and finance of decentralised CHP.	With commissioner, typically local governments, housing corporations, DSOs, some private households and companies.	Positive	Lack of information, many barriers from other stakeholders indirectly harm contracting parties.

Distribution system operator			Depending on ownership	
Owned by municipality (public utility company; PUC)	Public utility companies operate part of local heating grids.	Local government, end users, housing corporations, centralised energy company. Local government owns PUC.	Positive	Lack of information, concessions preventing ownership of all local heating grids.
Owned by RWE/E.ON	Large energy companies possess long-term concessions to operate a part of the heating grids.	Idem. However, energy company operates grid via concessions, not the PUC.	Negative	Potential ending of concession contracts. Loss of local heating grid monopolies.

Viétor *et al.*

Viétor *et al.* *Energy, Sustainability and Society* 2015 5:5, doi:10.1186/s13705-015-0033-0

Perceived Drivers and Barriers by Stakeholders

Several drivers and barriers regarding uptake and integration of decentralised CHP can be identified. They can be categorised as: investments, regional characteristics, information deficit and lack of awareness, policy-related barriers, district heating and lack of market services. They will now be discussed in more detail.

Investments

Decentralised CHP units are still considered relatively *expensive* (especially upfront investments). Payback periods are considered quite long. Linked to this, another barrier mentioned is the lack of knowledge about business models in relation to contracting or leasing. Return on investment is considered rather low. This is related to legal framework schemes changing frequently and the feed-in tariff for electricity generation, which is becoming unpredictable. Interviewees often draw comparison to the uptake of solar PV in the country. They claim that 'Feed-in tariffs for PV-electricity decreased dramatically in recent years and continue to do so on a monthly basis'. This creates investment uncertainties. Nonetheless, most interviewees believe that current policy and funding conditions are beneficial in respect of the uptake and integration of decentralised CHP.

Among the driving factors mentioned by interviewees increasing energy prices were frequently mentioned, which were linked to the aspiration of establishing 'energy autarkies' and not being dependent anymore on the large-scale energy producers. This idea is currently receiving more attention from different end-user groups. Another driving factor (but linked to the former) that could spur the development of decentralised CHP is the increasing popularity among citizens and local business companies for making local investments. The interviewees observe a trend that 'goes local'. People wish to invest in projects or organisations they can trust and relate to; often in close proximity to their homes. Decentralised CHP plants would offer such investment opportunities for local communities. Citizens' local cooperatives or local business in collaboration with private investors jointly provide the necessary financial means to collectively design, build and operate (renewable) energy production plants, for example wind farms. This example makes clear that our interviewees believe that financial barriers could be potentially be resolved by collective local investments, especially if they are based on popular support by local communities, notably local*Bürgerenergiegenossenschaften* (authors' translation: citizens' energy cooperatives).

Regional Characteristics

The Ruhr Valley has several specific characteristics that contribute to this area being rather unique. It has a large stock of old buildings, many having a status as a historical monument. Our interviewees view this as a positive reason for installing decentralised CHP units. They state that it is often more economic than retrofitting old buildings (by applying thermal insulation). In the case of conservation of monuments (especially in old fashioned mining settlement sites), basic thermal insulation cannot simply be applied since it would damage the fabric of the building and spoil the appearance. Moreover, our interviewees state that there is a good potential to use decentralised CHP units in urban districts that are not connected to district heating grids.

A second condition that could favour the uptake of decentralised CHP technology is the regional high-population density and having several big cities sited in close proximity. These conditions allow for a high-heat production potential in the region at close proximity to sites of high heat (and electricity) demand by end users. The important issue here is combining sufficient regional and local heat demand, since CHP plants can only operate economically and efficiently if a significant demand for heat can be ensured and transportation infrastructure of heat is not too long when taking into account potential heat loss.

Third, from an infrastructural point of view, the Ruhr Valley is characterised by an extensive existing district heating grid through which heat is already distributed to many city districts. This grid has however also been listed by the interviewees among the barriers to further uptake and development of decentralised CHP since the current district heating grid supplies sufficient amounts of heat already (from centralised industrial heat production plants). However, the existing grid offers important preconditions for low-carbon heat supply. The interviewees advocate exploring ways to create interconnections between the existing district heating grids and decentralised CHP units. In relation to this claim, one of our interviewees stated: 'It is not the grid that hinders the energy system transition in Germany. It is the energy production utilities that are centralized, large-sized and inflexible plants. Moreover, there is more potential for decentralised units to be integrated (better) into the current district heating system'.

The fourth regional-specific factor, and barrier, is the difficult financial situation in which the majority of the municipalities in the Ruhr Valley now find themselves. They have limited financial means and therefore are not in the position to allocate sufficient budget to promote the uptake and integration of decentralised CHP. Consequently, it is not possible to implement substantial earmarked policy support measures to convince local building owners in making appropriate upfront investments. Next to awareness-raising campaigns and other 'soft policies' local - non-governmental - building owners are left to finance investments in CHP themselves, unless they are able to attract other financial sources for making investments. Another negative consequence from having limited budgetary means is that municipalities have only a limited opportunity to become local 'launching customers'.

Lack of Awareness and Information Deficit

All interviewees mention the important influence of information provision and awareness raising on the uptake and integration of decentralised CHP in the Ruhr Valley by local stakeholders. They agree, however, that decentralised CHP technology is not familiar among its potential target groups (i.e. private homeowners, housing corporations and companies looking for office space). CHP units are considered as 'not visible' by end users, as they are located in places where people do not notice them, for example in cellars. Interestingly, technological equipment suppliers and service companies (such as those responsible for installation) are often not familiar with the technology. Interviewees claim that it is, to a large extent, this information deficit that is responsible for the lack of decentralised CHP usage in the Ruhr Valley. Several interviewees state that promoting this technology is not easy as the concept is perceived as 'rather difficult to understand'. Moreover, the technology is considered as rather 'complex', notably by those in society who do not engage with energy technology regularly.

Another barrier mentioned by respondents concerns the habits and behaviour of end users. In particular, it is difficult to change habits. However, once end users are familiar with using an energy technology, it is difficult to convince them to switch to using a new (energy) technology (e.g. decentralised CHP) since it requires moving away from a proven technology the user is familiar with. The interviewees suggest that it is possibly easier to convince target groups, when promoting the technology, by mentioning the particular benefits of CHP that would appeal to them. However, this strategy would require data about the target group's heat profiles and that decentralised CHP energy production can be tailored to their specific needs. Unfortunately, there appears to be little information available on individual end-user heat profiles.

Interviewees mentioned pilot projects as a positive driving factor to encourage the uptake and development of decentralised CHP. Pilot projects receive a lot of public attention. It creates awareness and helps familiarise potential adopters of the technology. If the results of a pilot project are positive, then a basis for trust and further experimentation is likely to have been created. Decentralised CHP then becomes known to the potential end users as a more or less 'proven technology'. According to the interviewees, trust in the technology and its installers is of particular importance. It is not unreasonable to assume that one would only invest and adopt, or participate in collectively buying, a decentralised CHP unit if one is convinced that this technology is proven and has long-term added value both financially and environmentally.

Policy Barriers

Current Federal and State policy frameworks that focus on supporting decentralised CHP technology are viewed by the interviewees as having been well formulated. Essentially, they are considered to create sufficient framework conditions to foster uptake. Several funding instruments are deemed useful and considered as positive driving factors. However, some interviewees believe that the amount of funding offered (for example by feed-in tariffs and subsidies through grants) is not sufficient to promote the uptake. This response seems to be related to the situation in many municipalities in the Ruhr Valley which are suffering from stretched financial budgets (related to the economic crisis). For this reason, municipalities find it difficult to co-fund investments with potential local adopters.

A second barrier mentioned by the interviewees relates to the way in which climate policies have been formulated. Interviewees considered these policies overly ambitious with too long a time horizon as an often cited criticism. As one interviews states: 'Thinking in "big steps" is not appropriate if one strives to really attain climate mitigation and energy transition goals set'. 'Setting policy goals that have to be attained in fifteen, twenty or even forty years fail to create responsibilities as politicians and public officials who are currently in power will – most likely - not be in power anymore once these policy objectives have been or should have been met.' Another interviewee comments underline this point: 'Policymakers hide behind long term goals so that they cannot be held responsible once short term policy goals have not been met'. Hence, in general, climate mitigation policy goals are seen as unrealistic, lacking attention to situational settings, and henceforth fail to address realistic, feasible opportunities to promote the use of decentralised CHP.

Third, complex policy framework conditions are seen as a hindrance factor *vis-à-vis* the uptake and integration of decentralised CHP. For instance, one interviewee states: 'Once one applies for a financial support scheme one needs to fill in and register a decentralised CHP unit in order to be entitled to receive payments.' This is perceived as a difficult and time-consuming task. Thus, it is considered challenging to comply with all the necessary formalities. This situation applies to both private homeowners and housing corporations. It is with the latter where the greatest potential for adopting decentralised CHP units lies. These corporations are in a position where they can sell both heat and electricity directly to their tenants. However, the contemporary legal framework hinders the housing corporations in becoming energy producers. Opinion is divided between stakeholder groups about whether or not this regulation is the main barrier that prevents housing corporations from installing decentralised CHP units.

A fourth policy-related barrier is the strong coal and gas lobby in the Ruhr area. Policymakers and lobbyists are considered to have very close working relations. The influence that is exerted by the coal and gas lobby on policy making is seen as a strong hindrance to policymaking *vis-à-vis* supporting the uptake and integration of decentralised power production in general and renewable energy-based technologies in particular.

A fifth policy barrier mentioned is the 2013 national elections.[h] Prior to the 2013 elections, interviewees acknowledge it is difficult to foresee which parties will be involved in the new government coalition, and whether the current policies and programs will be maintained. Interviewees claim that it is necessary to wait and see whether developments after the elections will affect current policy framework conditions, especially the long-term goals and the decentralised CHP support policies. As a consequence, investment decisions are postponed until after the elections have taken place and a new government coalition is in power and has published its policy strategies.

[h]Data were collected in the weeks prior to the September 2013 elections.

District Heating

As mentioned above, the existing district heating grid is seen as a barrier to the uptake and integration of decentralised CHP. Expectations are that several (notably gas-fired) power plants will have to shut down in the near future. This already resulted in widespread uncertainties among stakeholders regarding the continuation of heat supply to the district heating grid. One interviewee underlines this by stating that 'A difficult situation needed to be overcome when it turned out that the coal power plant "Datteln 4" with a capacity of 1,050 MW would not start operating in time (if at all). Due to planning mistakes the power plant could not start feeding heat into the district heating grid. As a consequence an alternative solution had to be found'. This event led to strong public protests. Such situations - creating tensions to the current hegemonic energy regime in the region - can consequently spur the development of decentralised energy production (including by small- and medium-sized CHP units) to ensure the continued availability of local heat production. The emphasis on 'decentralised and local' is considered important since it creates independence from the large-scale heat producers from whom security of supply cannot be expected anymore which represents an undermining of trust considered as a factor in the stability of the dominant energy regime.

The role of municipal public utility companies (PUCs) can be seen as a potential influential factor in this respect. In cities that own PUCs, the interviewees observe a large potential to install decentralised CHP units: 'Municipalities could potentially start projects in which public grids connect to decentralised CHP units and their end users.' However, ownership of district heating grids by PUCs is not found in all cities in the Ruhr Valley. In some cities, energy companies like RWE and E.ON have district heating grid concessions. This means that municipalities which want to support the integration of decentralised CHP in district heating systems will be dependent on the willingness of energy companies to see this scale of technology as part of their technology mix. This dependency of the municipalities on a stakeholder which has not shown enthusiasm about decentralised systems is viewed as a substantial barrier by the interviewees.

Lack of Market Services

A lack of companies providing services regarding installation and maintenance of decentralised CHP units is perceived as an important barrier by the interviewees. Two reasons were advanced to explain this view. First, installation workers lack proper instructions on how to install and maintain decentralised CHP units. Without these instructions, they simply cannot offer these services to customers, particularly in a manner that is likely to create trust. However, producers of CHP units and installers do offer courses and training sessions about installation and maintenance of the decentralised units. The difficulty is that such courses need to fit into the day-to-day working of these (typically small-sized) companies. Usually, they have little time or spare staff available (if at all) to participate in these courses, particularly if they are offered during working hours. A related problem in this regard is that this type of small-sized service companies lacks the capacity to specialise in developing services regarding installation and maintenance of decentralised CHP units. This results in a situation where very few (specialised) people need to cover an increasing range of specialised services related to particular technological products.

Results of Systemic Analysis Using MLP

The stakeholders' analysis presented an overview of barriers that can be seen as regime resistance forces. Major identified barriers relate to (i) lack of financial means for making investments, (ii) lack of awareness regarding the benefits of decentralised CHP to potential users, (iii) the presence of centralised district heating systems, (iv) lack of sufficient policies supporting diffusion of decentralised CHP units, (v) legal stipulations from social housing policies that prevent housing cooperatives from becoming energy producers, (vi) district heating systems owned by public and private owners (via concessions contracts), (vii) lack of skilled service-providing companies and (viii) vested interests by the coal and gas industry. Many of the identified barriers seem interrelated, especially those related to policies promoting decentralised CHP units and the required investments and training of skilled personnel. In MLP, the identified barriers fall within the following regime forces: energy markets, user preferences, (sectoral) policies and industry. We will now present the results of the MLP analysis, using the

Decentralised Combined Heat and Power in the German Ruhr Valley...

91

MLP categorization of the macro, meso and micro levels.

Macro Level: Landscape Factors

Although there are many regime barriers (i.e. the problems presented in the previous section), there are also many ('landscape') developments that could potentially offer windows of opportunity to encourage the uptake of decentralised CHP technology. The security of energy supply is no longer taken for granted. Recent problems with heat supply and expectations of rising energy prices can be considered to have played a role in local communities being stimulated to organise their own energy supply with the emphasis on local provision, particularly at the district level. This can be seen as part of a 'decentralisation' trend in Germany [59] in which citizens want to move away from a large centralised organisation of utilities. In that sense, decentralised CHP technology offers a promising alternative, as both heat and power can be produced efficiently at the district level. Moreover, it also allows for the use of renewable energy sources (which appeals to citizens who prefer 'green' energy sources to fossil ones). Connecting to the district heating systems could offer interesting opportunities for upscaling decentralised CHP use (e.g. from the building level to the district level). However, this requires that concession holders of the district heating grid allow for such use and do not only use it but also sell excess heat to industry and the large fossil-fuelled power plants. The ending of district heating systems' exploitation concessions in 2015 offers a window of opportunity for municipalities, local communities and perhaps citizen's energy cooperatives to consider decentralised CHP. Policy frameworks, in particular the *Energiewende*program, can influence the policy agendas of NRW and the Ruhr Valley municipalities including ownership of the heating grids. Nonetheless, another landscape event, the economic crisis, has had a direct negative financial impact on the capability of governments to allocate budgets for investments in the niche development of decentralised CHP units.

Meso Level: Regime Factors

Currently, potential users (particularly social housing corporations and private homeowners living in multistorey buildings) are hardly aware of the technology's benefits. When users are aware of the benefits, they appear not to be willing to make the investment required to install decentralised CHP units. This latter situation might be linked to the decision-making practices of building owners. First, private homeowners need collective action in order to decide on investments and plans towards installing CHP units. This is time consuming with high-transaction costs. Second, regulations do not allow housing corporations to become energy producers. In addition, decentralised CHP does not seem to match with current user preferences and habits concerning (common) energy consumptive practices. Without social housing policy changing, an important stakeholder is left without institutional allowance to even start using (medium-sized) decentralised CHP units. This situation creates the impression that there is, as yet, little market demand. The installation sector being not willing to invest in specialising in decentralised CHP technology and therefore lacking service provision to accommodate potential users of the technology is also problematic.

The situation with these two stakeholder groups' lack of enthusiasm for decentralised CHP can be linked to the absence of effective support policies, in particular policy instruments to support potential users in making upfront investments. As a consequence, potential users adopt a 'wait and see' strategy and delay making investments until sufficient, trustworthy government programs are in place. As a consequence, current conditions for creating a decentralised CHP (niche) market in the Ruhr Valley are limited. It seems that decentralised CHP still represents a niche in an infant development stage: take off and diffusion into larger market segments is currently not taking place.

Micro Level: Niche Development

Within the Ruhr Valley, a protected space in which room for experimentation with decentralised CHP is created is the 'Innovation City Ruhr' in the city of Bottrop. The latter was selected by the State of NRW to become a 'climate friendly model city' in the Ruhr Area. The aim is to transform a city district into an energy efficient area, which includes the construction of a decentralised CHP unit based on renewable energy sources, and in so doing provides a best practice example for the larger Ruhr Valley region.

In addition, the policies designed by the nine Ruhr Valley municipalities to support decentralised CHP uptake show a wide scope of actions of how this 'niche' can be further developed. Actions planned by Ruhr area municipalities differ in various ways. Whereas some municipalities only strive to install decentralised CHP in public buildings, others target local stakeholders, like households and commercial firms. Moreover, municipalities deploy different policy actions, such as constructing apartment block-heating installations, integration of decentralised CHP units in urban development planning and establishing contracts with private home owners for the provision of heat. Decentralised CHP is also seen as a means to feed in decentralised-produced heat into the existing district heating grids. Studies regarding the technical and economic potential or feasibility of decentralised CHP have been prepared in municipalities, and plans for constructing decentralised CHP units are prepared. The ways the city governments address the uptake and integration of decentralised CHP show little coherence: there is little alignment in the formulation of goals and coordination policy actions between Ruhr Valley cities. From a systemic - strategic niche management - perspective, the proposed actions look somewhat promising on the one hand, but rather uncoordinated and little aligned on the other hand. Moreover, there seems not to be sufficient shielding and nurturing of the 'decentralised CHP niche'. The same can be said in respect of the proposed actions to destabilise the incumbent regime structure. Nonetheless, as mentioned before, several landscape events are creating tensions in the regime which could create momentum for further niche development. In the end, however, municipalities cannot bring about systemic change alone. Although cities in the Ruhr Area would like to contribute to climate protection goals with their own utility companies and the possession of the electricity grid, they are depending on other stakeholders, like energy companies and distributed system operators. Many of the district heating concession contracts will end in 2015 which opens up opportunities for regime change. Energy companies are said (according to our interviewees) to use gaps in the energy anti-monopoly regulations to hamper this development. Therefore, a change of regulatory schemes is deemed a necessary requirement to create a level playing field [60].

DISCUSSION

The *Energiewende* does not necessarily only need to rely on supporting green electricity production. Actors in urban environments like the Ruhr Valley could also embrace harvesting the potential of energy efficiency goals, *in casu* adopting decentralised CHP technology. The results, however, show that many barriers have been identified that block the uptake and integration of this technology in practice.

The stakeholder analysis proved useful in giving us more insight into the 'agency' factor of the Ruhr Valley energy system. In line with [38]-[40], we feel that providing more insight into agency (in our case, by conducting stakeholder analysis prior to using/applying MLP) provides/adds complementary functions in a methodological sense. It contributes to identifying regime barriers and interlevel system dynamics. We leave it to scholars in the field of sustainable transitions to judge whether this can be concerned a methodological 'novel insight'.

Furthermore, we feel that paying attention to specific regional characteristics of the Ruhr Valley is worthwhile as well. Without understanding the industrial character, the availability of district heating infrastructures (and its ownership contracts via concessions), the influence of the gas and coal lobby, the large-scale of dwellings owned by social housing corporations and the decentralised governments who can afford little budget to formulate and implement significant niche supporting programs, it is difficult to understand why the Ruhr Valley's energy system is not responding positively to the uptake and integration of decentralised CHP. Hence, in line with [50], we argue that paying special attention to geographically bound comparative (dis-)advantageous circumstances matters in understanding energy system dynamics. Furthermore, we agree with[48],[49] that policies also matter and therefore deserve sufficient analytical attention. If policies that intend to support niche developments are non-aligned and are not coordinated with sectoral policies (e.g. social housing policies addressing housing corporations' entitlements in not becoming energy producers), niche development (e.g. uptake of decentralised CHP technology) is not expected to take off [52]. In addition, in the absence of using 'systemic instruments' that exceed local authorities' action plans and jurisdictions, this is not expected to happen either [48]. In conclusion, we agree with Geels [52] that the understanding of power relations at the regime level is critical to the further understanding of greening energy systems. In our opinion, this especially applies to urban, industrial and energy-intensive regions.

Next to addressing the use of MLP and reflecting on the added value of the theoretical contributions to elaborate on MLP, TM and SNM, as mentioned in the academic literature (see the paragraph in the 'Theoretical framework: using the multilevel perspective to analyse regional energy transitions' section), we also want to address the commonalities that our analysis show with results from empirical studies addressing barriers that impede the adoption of (green) energy innovations in city districts. They concern difficulties experienced when trying to have such innovations adopted in existing urban areas, such as the lack of finance, lack of trust and little effective (local) government policies to persuade public (social housing associations) [61],[62] and private homeowners [63] to adopt measures to increase the energy efficiency of dwellings. As shown in the Ruhr Valley case study, the lack of market demand and the lack of incentives from the regulatory framework fail to stimulate change in the energy regime. The situation resembles the status quo in (existing) urban contexts in other Western Europe with a lack of diffusion of 'green' energy innovations (see e.g. [64]). Learning from these experiences and taking these lessons into account can be useful for those interested in studying and formulating policies to stimulate adoption of these kinds of innovations (notably addressing energy efficiency), which is now seen as a key challenge in light of the German *Energiewende*[22].

Generalisation of the results from this study is difficult since we focus on only one case (the Ruhr Valley region). The study, however, can be replicated in other regions. These regions should, however (at least) match some of the main characteristics of the Ruhr Valley region: highly industrialised, energy intensive and urbanised. Studying the results and system innovation indicators of theoretical interest in a set of multiple (comparable) cases would allow for more systematic research and would contribute to the academic literature.

CONCLUSIONS

Energy-intensive regions and energy efficiency in buildings are considered to have taken a backseat in Germany's *Energiewende* program. Recently, the importance of these issues has received scholarly attention [23]. In this paper, we contribute to this agenda. The main question in this paper is: *Which factors block the uptake and integration of decentralised CHP in the German Ruhr area's energy system?*

Following a stakeholder analysis, and a systemic analysis using MLP, the results show that the incumbent regime provides substantial barriers that prevent large-scale uptake and integration of decentralised CHP. These barriers relate to lack of financial means for making investments, unawareness and information deficit regarding the benefits of decentralised CHP to potential users, the presence of centralised district heating systems, lack of sufficient policies supporting diffusion of decentralised CHP units, legal stipulations from social housing policies that prevent housing cooperatives from becoming energy producers, district heating systems owned by public and private owners (via concessions contracts), lack of skilled service-providing companies and vested interests by the coal and gas industry. Moreover, many of the mentioned factors seem interrelated, especially those linked to policy and finance for upfront investments.

Although the study presented in this paper was conducted in a region with rather unique characters, it would be worthwhile to further the research agenda on the uptake and integration - and more in general: niche development - of decentralised CHP or other decentralised energy production technologies - in regions. Future academic work in this field would particularly be useful if comparative regional studies are conducted. Our paper has also demonstrated the added value of conducting stakeholder analysis as a complementary method to the use of MLP, notably as a way that agency can be explored. Moreover, we argue that conducting a stakeholder analysis prior to studying sustainable transitions in case study research designs using the more commonly used theoretical frameworks and heuristic tools in this disciplinary field (e.g. MLP or SNM) contributes in a methodological sense to the way transition studies address agency. For this reason, we encourage future case studies on sustainable transitions to adopt stakeholder analysis methodology.

AUTHORS' CONTRIBUTIONS

All authors were involved in drafting the manuscript. All authors read and approved the final manuscript.

ACKNOWLEDGEMENTS

The authors would like to thank the Wuppertal Institute for Climate, Environment and Energy and particularly Dr. Johannes Venjakob for his contribution. The institute hosted the main researcher and supported the identification of the stakeholder network and the organisation of interviews. We also like to thank three independent reviewers for assessing the manuscript and providing us with suggestions for improvement.

REFERENCES

1. Jacobssen S, Johnson A (2000) The diffusion of renewable energy technology: an analytical framework and key issues for research. Energy Policy 28:625-640

2. Loorbach D, Verbong G (2012) Governing the Energy Transition, Reality, Illusion or Necessity. Routledge, New York, London.

3. Westner G, Madlener R (2011) Development of cogeneration in Germany. A mean-variance portfolio analysis of individual technology's prospects in view of the new regulatory framework. Energy 36:5301-5313

4. Agentur für Erneuerbare Energien (2012) Deutschlands Informationsportal für Erneuerbare Energien. Berlin.

5. Ministerium für Klimaschutz, Umwelt, Landwirtschaft, Natur- und Verbraucherschutz des Landes NRW (2013) Kraft-Wärme-Kopplung. Impulse für Energiewende "Made in NRW". Düsseldorf https://broschueren.nordrheinwestfalendirekt.de/broschuerenservice/energieagentur/kraft-waerme-kopplung/1550. Accessed on 16 July 2013

6. (2012) Gemeinsam für die kommunale Energiewende. Auf dem Weg in eine klimafreundliche Zukunft, Berlin.

7. (2011) Deutschlands Energiewende. Ein Gemeinschaftswerk für die Zukunft, Berlin.

8. Hoppe T, Coenen F (2011) Creating an analytical framework for local sustainability performance: a Dutch case study. Local Environ 16(3):229-250

9. Truffer B, Coenen L (2012) Environmental innovation and sustainability transitions in regional studies. Reg Stud 46(1):1-21

10. EU Tech Energie & Management (2008) NRW-Klima2020 – Beitrag Nordrhein-Westfalens zur Erreichung des nationalen Klimaschutzziels. Aachen

11. Rip A, Kemp R (1998) Technological change. In: Rayner S, Malone EL (eds) Human Choice and Climate Change, Battelle, Columbus. pp 327-399

12. Geels F (2002) Technological transitions as evolutionary reconfiguration processes: a multi-level perspective and a case study. Res Policy 31:1257-1274

13. Regionalverband Ruhr (2013) Regionalverband Ruhr – Aufgaben und Verbandsgebiet., Essen http://www.metropoleruhr.de/regionalverband-ruhr/ueber-uns/gebiet-aufgaben.html. Accessed on 28 May 2013

14. Ruhrgebiet Regionalkunde (2013) Die administrative Gliederung des Ruhrgebietes., Essen http://www.ruhrgebiet-regionalkunde.de/grundlagen_und_anfaenge/kohle/kohle.php?p=2 = 0 Accesed on 28 May 2013

15. Geographie Infothek (2012) Infoblatt Strukturwandel im Ruhrgebiet. Stuttgart http://www2.klett.de/sixcms/list.php?page=geo_infothek&miniinfothek=&node=Ruhrgebiet&article=Infoblatt+Strukturwandel+im+Ruhrgebiet. Accessed on 22 May 2012

16. Beggs C (2011) Energy: Management, Supply and Conservation. Spon Press, London/New York.

17. Dena – German Energy Agency (2013) Combined heat and power generation. Berlin http://www.dena.de/en/projects/energy-systems/combined-heat-and-power-generation.html Accessed on 26 June 2013

18. Dena – German Energy Agency (2013): Kraft-Wärme-Kopplung. Berlin   http://www.effiziente-energiesysteme.de/themen/kraft-waerme-kopplung/herausforderungen.html Accessed on 26 June 2013

19. (2009) Langfristszenarien und Strategien für den Ausbau erneuerbarer Energien in Deutschland. Leitszenario, Berlin.

20. Eikmeier B, Klobasa M, Toro F, Menzler G (2011) Potenzialerhebung von Kraft-Wärme-Kopplung in Nordrhein-Westfalen.

21. BET – Büro für Energiewirtschaft und technische Planung GmbH (2013) Perspektiven der Fernwärme im Ruhrgebiet bis 2050. Aachen

22. Gawel E, Lehmann P, Korte K, Strunz S, Bovet J, Köck W, Massier P, Löschel A, Schober D, Ohlhorst D, Tews K, Schreurs M, Reeg M, Wassermann S (2014) The future of the energy transition in Germany. Energy Sustainability Soc 2014(4):15

23. Kemp R (1994) Technology and the transition to environmental sustainability. Futures 26:1023-1046

24. Geels FW, Schot J (2010) The dynamics of sociotechnical transitions – a sociotechnical perspective. In: Grin J, Rotmans J, Schot J (eds) Transitions to Sustainable Development, Routledge, New York. pp 9-101

25. Geels F (2004) From sectoral systems of innovation to socio-technical systems. Insights about dynamics and change from sociology and institutional theory. Res Policy 33:897-920

26. Markard J, Raven R, Truffer B (2012) Sustainability transitions: an emerging field of research and its prospects. Res Policy 41(6):955-967

27. Smith A, Stirling A, Berkhout F (2005) The governance of sustainable sociotechnical transitions. Res Policy 34:1491-1510

28. Unruh GC (2000) Understanding carbon lock-in. Energy Policy 28:817-830

29. Markard J (2011) Transformation of infrastructures: sector characteristics and implications for fundamental change. J Infrastructure Syst (ASCE) 17:107-117

30. Verbong G, Geels F (2007) The ongoing energy transition: lessons from a socio-technical, multi-level analysis of the Dutch electricity system (1960–2004). Energy Policy 35(2):1025-1037

31. Smith A, Raven R (2012) What is protective space? Reconsidering niches in transitions to sustainability. Res Policy 41:1025-1036

32. Kemp R, Schot J, Hoogma R (1998) Regime shifts to sustainability through processes of niche formation: the approach of strategic niche management. Tech Anal Strat Manag 10(2):175-198

33. Raven RPJM (2005) Strategic niche management for biomass: a comparative study on the experimental introduction of bioenergy technologies in the Netherlands and Denmark. PhD thesis, TU/E Eindhoven

34. Rotmans J, Kemp R, van Asselt M, Geels F, Verbong G, Molendijk K (2000) Transities & transitiemanagement. De casus van een emissiearme energievoorziening, Rotterdam.

35. Loorbach DA (2007) Transition management: new mode of governance for sustainable development. PhD thesis. Erasmus University, Rottredam

36. Geels FW, Kemp R (2005) Transitions, transformations and reproduction: dynamics in socio-technical systems. Paper presented at the DRUID Tenth Anniversary Summer Conference, Copenhagen Denmark

37. Shove E, Walker G (2007) CAUTION! Transitions ahead: politics, practice and sustainable transition management. Environ Plann A 39:763-770

38. Van den Bergh JC, Truffer B, Kallis G (2011) Environmental innovation and societal transitions: introduction and overview. Environ Innov Soc Trans 1(1):1-23

39. Raven RPJM, Verbong GPJ, Schilpzand WF, Witkamp MJ (2011) Translation mechanisms in socio-technical niches: a case study of Dutch river management. Tech Anal Strat Manag 23:1063-1078

40. Farla J, Markard J, Raven R, Coenen L (2012) Sustainability transitions in the making: a closer look at actors, strategies and resources. Technol Forecast Soc Chang 79(6):991-998

41. Markard J, Truffer B (2008) Actor-oriented analysis of innovation systems: exploring micro-meso level linkages in the case of stationary fuel cells. Tech Anal Strat Manag 20:443-464

42. Grin J (2010) Understanding transitions from a governance perspective. In: Grin J, Rotmans J, Schot J (eds) Transition to Sustainable Development, Routledge, New York/London. pp

221-337

43. Avelino F, Rotmans J (2009) Power in transition: an interdisciplinary framework to study power in relation to structural change. Eur J Soc Theory 12(4):543-569

44. Avelino F (2011) Power in transition: empowering discourses on sustainability transitions. PhD thesis, Erasmus University, Rotterdam

45. Shove E, Walker G (2010) Governing transitions in the sustainability of everyday life. Res Policy 39:471-476

46. Seyfang G, Smith A (2007) Grassroots innovations for sustainable development: towards a new research and policy agenda. Environ Polit 16(4):584-603

47. Arentsen M, Bellekom S (2014) Power to the people: local energy initiatives as seedbeds of innovation? Energy Sustainability Soc 4:2

48. Smits R, Kuhlmann S (2004) The rise of systemic instruments in innovation policy. Int J Foresight Innov Policy 1(1):4-32

49. Weber M, Rohracher H (2012) Legitimizing research, technology and innovation policies for transformative change. Res Policy 41:1037-1047

50. Coenen L, Benneworth P, Truffer B (2012) Towards a spatial perspective on sustainability transitions. Res Policy 41:968-979

51. Geels FW (2011) The multi-level perspective on sustainability transitions: responses to seven criticisms. Environ Innov Soc Trans 1(1):24-40

52. Geels FW (2014) Regime resistance against low-carbon transitions: introducing politics and power into the multi-level perspective. Theory, Culture & Society doi:10.1177/0263276414531627

53. Bulkeley H, Betsill M (2003) Cities and Climate Change: Urban Sustainability and Global Environmental Governance. Routledge, New York London.

54. Bulkeley H (2013) Cities and Climate Change. Routledge, New York London.

55. Hoppe T, van den Berg MM, Coenen FHJM (2014) Reflections on the uptake of climate change policies by local governments: facing the challenges of mitigation and adaptation. Energy

Sustainability Soc 4:8

56. Hoppe T, van Bueren E (2014) Governing the climate challenge and energy transition in cities. Energy, Sustainability and Society, Thematic Series., http://www.energsustainsoc.com/series/GCCE

57. Gerring J (2007) Case Study Research. Principles and Practices, Cambridge.

58. Varvasovszky Z, Brugha R (2000) How to do (or not to do) ... A stakeholder analysis. Health Policy Plan 15:338-345

59. Oteman M, Wiering M, Helderman JK (2014) The institutional space of community initiatives for renewable energy: a comparative case study of the Netherlands, Germany and Denmark. Energy Sustainability Soc 4(1):11

60. Berlo K, Wagner O (2013) Harter Gegenwind bei der Rekommunalisierung. AKP 4:22-23

61. Hoppe T, Bressers H, Lulofs K (2010) Energy Conservation in Dutch Housing Renovation Projects. In: Martens P, Chiang CT (eds) The Social and Behavioural Aspects of Climate Change, Greenleaf, Sheffield. pp 68-95

62. Hoppe T (2012) Adoption of innovative energy systems in social housing: lessons from eight large-scale renovation projects in The Netherlands. Energy Policy 51:791-801

63. Murphy L, Meijer F, Visscher H (2012) A qualitative evaluation of policy instruments used to improve energy performance of existing private dwellings in the Netherlands. Energy Policy 45:459-468

64. Faber A, Hoppe T (2013) Co-constructing a sustainable built environment in the Netherlands—Dynamics and opportunities in an environmental sectoral innovation system. Energy Policy 52:628-638

Origin and Transport of Pore Fluids in the Nankai Accretionary Prism Inferred from Chemical and Isotopic Compositions of Pore Water at Cold Seep Sites off Kumano

Tomohiro Toki[1], Ryosaku Higa[1], Akira Ijiri[2], Urumu Tsunogai[3, 5], and Juichiro Ashi[4]

[1]Department of Chemistry, Biology and Marine Science, Faculty of Science, University of the Ryukyus, 1 Senbaru, Nishihara 903-0213, Okinawa, Japan

[2]Kochi Institute for Core Sample Research, JAMSTEC, B200 Monobe, Nankoku 783-8502, Japan

[3]Earth and Planetary System Science, Faculty of Science, Hokkaido University, N10 W8, Kita-ku, Sapporo 060-0810, Hokkaido, Japan

[4]Department of Ocean Floor Geoscience, Atmosphere and Ocean Research Institute, The University of Tokyo, 5-1-5 Kashiwanoha, Kashiwa 277-8568, Chiba, Japan

[5]Current address: Graduate School of Environmental Studies, Nagoya University, Furo-cho, Chikusa-ku, Nagoya 464-8601, Japan

ABSTRACT

We used push corers during manned submersible dives to obtain sediment samples of up to 30 cm from the subseafloor at the Oomine Ridge. The concentrations of B in pore water extracted from the sediment samples from cold seep sites were higher than could be explained by organic matter decomposition, suggesting that the seepage fluid at the site was influenced by B derived from smectite-illite alteration, which occurs between 50°C and 160°C. Although the negative $^{18}O_{H2O}$ and D_{H2O} values of the pore fluids cannot be explained by freshwater derived from clay mineral dehydration (CMD), we considered the contribution of pore fluids in the shallow sediments of the accretionary prism, which showed negative $^{18}O_{H2O}$ and D_{H2O} values according to the results obtained during Integrated Ocean Drilling Program (IODP) Expeditions 315 and 316. We calculated the mixing ratios based on a four-end-member mixing model including freshwater derived from CMD, pore fluids in the shallow (SPF) accretionary prism sediment, seawater (SW), and freshwater derived from methane hydrate (MH) dissociation. However, the Oomine seep fluids were unable to be explained without four end members, suggesting that deep-sourced fluids in the accretionary prism influenced the seeping fluids from this area. This finding presents the first evidence of deep-sourced fluids at cold seep sites in the Oomine Ridge, indicating that a megasplay fault is a potential pathway for the deep-sourced fluids.

BACKGROUND

At cold seeps, pore fluids seep from sea-bottom sediments. These seepage fluids are generally enriched in CH_4 or H_2S, and chemosynthetic communities such as bacterial mats and *Calyptogena*, which use CH_4 or H_2S as an energy source, cluster on the seafloor at cold seep sites

(Paull et al. [1984]; Suess et al. [1985]). Cold seep sites have been observed in subduction zones and in passive margins worldwide, and the seepage fluids have been reported to have various sources (Suess et al. [1985]; Wallmann et al. [1997]; Aharon and Fu [2000]; Lein et al. [2000]; Naehr et al. [2000]; Greinert et al. [2002]). The Nankai Trough subduction zone is a convergent plate margin where the Philippine Sea plate is subducting below the Eurasian plate (Figure 1a). The surface sediments on the subducting plate have accreted on the landward slope of the Eurasian plate, forming an accretionary prism that consists of a toe, slope, and outer ridge. Within the slope sediments, megasplay faults branch from the plate-boundary interface and intersect the seafloor at the foot of the outer ridge (Park and Kodaira [2012]; Park et al. [2002]). One such intersection is the Oomine Ridge, where several bacterial mats have been observed on the seafloor (Toki et al.[2011], [2004]).

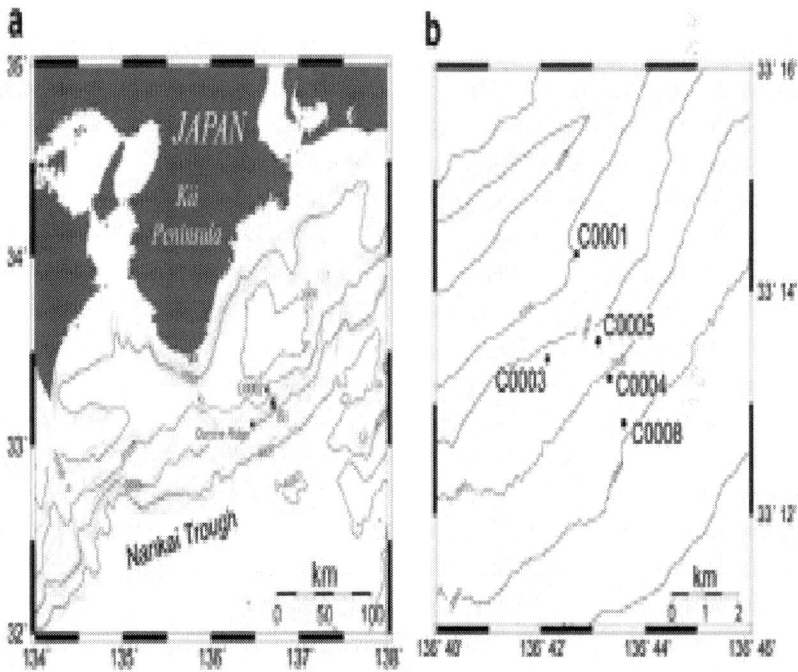

Figure 1: Bathymetric map and cross section of the Nankai Trough accretionary prism. (a) Bathymetric map of the Nankai Trough accretionary prism

off Kumano showing the location of the Oomine Ridge together with the locations of Sites C0001, C0002, C0003, C0004, C0005, and C0008, drilled during Integrated Ocean Drilling Program (IODP) Expeditions 314, 315, and 316. Contour is 100 m. A box indicates the area of Figure 1b. (b) Cross section of the Nankai Trough accretionary prism. Positions of drill Sites C0001, C0003, C0004, C0005, and C0008 are shown. Contour is 50 m.

Cl^- concentrations in the pore fluids at the bacterial mats are lower than that in seawater (SW) (Toki et al. [2004]). Deviations of the chemical and isotopic compositions of the pore fluids from those of SW have been attributed to the mixing of fluids with compositions differing from those of SW (Kastner et al. [1991]; Tsunogai et al. [2002]; Dählmann and de Lange [2003]; Mazurenko et al. [2003]; Toki et al. [2004]; Hiruta et al. [2009]). Since each of these fluid sources is characterized by specific O and H isotopic compositions (e.g., Kastner et al. [1991]). The seepage fluids on the Oomine Ridge have been inferred to originate from horizontally transported groundwater because the $^{18}O_{H2O}$ and D_{H2O} values of the pore fluids at the cold seep sites on the Oomine Ridge are in the range of those of groundwater in coastal northwestern Japan (Toki et al. [2004]).

In this study, we investigate the origin of the seepage fluids on the Oomine Ridge by examining concentrations of B. The level of B in seepage fluids is controlled mainly by interactions of sediment or rock with water, which depends on the temperature of the reaction (You and Gieskes [2001]). High B concentrations have been reported in pore fluids from mud volcanoes (Aloisi et al.[2004]; Teichert et al. [2005]; Haese et al. [2006]; Hensen et al. [2007]; Reitz et al. [2007]; Chao et al. [2011]). Proposed sources of high B concentrations in pore fluids are organic matter desorption at low temperature (Brumsack et al. [1992]; You et al. [1993b]) and smectite-illite alteration in the temperature range of 50°C and 160°C (You et al. [1996]). The high temperatures required for such alteration occur at great depths in sediments and rocks of the Earth's crust, depending on the thermal gradient in a given area. Generally, average thermal gradients are between 50°C/km and 60°C/km (Parsons and Sclater [1977]), and high-temperature environments of 150°C to 160°C are found at 2 to 3 km below the seafloor. Thus, by examining B concentrations, the contribution of deep-sourced fluids to pore fluids can be investigated (Martin et al. [1996]; Aloisi et al. [2004]; Haese et al. [2006]). Using a submersible, we collected sediment samples from sediment depths of up to 30 cm at cold seep sites on the Oomine Ridge, and we evaluated the chemical and isotopic compositions, especially the B concentrations, of pore fluids extracted from the sediments. Then, we inferred the origins of the seepage fluids from these chemical and isotopic compositions.

METHODS

Sampling

During cruise YK06-03 Leg 2 in May 2006 and cruise YK08-04 in April 2008 of the support ship *Yokosuka* of the Japan Agency for Marine-Earth Science and Technology (JAMSTEC), dive investigations were conducted by the JAMSTEC-manned submersible *Shinkai6500* on the Oomine Ridge (Figure 1). Sampling point locations and descriptions of dive 949 (YK06-03 Leg 2) and dive 1062 (YK08-04) are shown in Table 1.

Table 1: Location of sampling points during the YK06-03 nd YK08-04 cruises of the tender *Yokosuka*

Cruise	Date	Sample number	Latitude	Longitude	Depth (m)	Description
YK06-03	6 May 2006	D949 C1	33° 7.3283 N	136° 28.7705 E	2,519	Inside of a bacterial mat
	6 May 2006	D949 C3	33° 7.2253 N	136° 28.6672 E	2,533	Inside of a different bacterial mat

YK08-04	6 April 2008	D1062 C1	33° 7.3481 N	136° 28.7341 E	2,528		Inside of a bacterial mat
	6 April 2008	D1062 C2	33° 7.3481 N	136° 28.7301 E	2,531		Outside of a bacterial mat
	6 April 2008	D1062 C3	33° 7.2348 N	136° 28.5974 E	2,530		Inside of a different bacterial mat
	6 April 2008	D1062 C4	33° 7.2348 N	136° 28.5974 E	2,530		Outside of a different bacterial mat
	6 April 2008	D1062 C5	33° 7.2348 N	136° 28.5974 E	2,530		Inside of the bacterial mat samples in D1062 C3

Toki et al.

Toki et al. Earth, Planets and Space 2014 66:137, doi:10.1186/s40623-014-0137-3

Sediment cores up to about 30 cm long were collected from cold seep sites on the Oomine Ridge with MBARI-type push corers (http://www.mbari.org/dmo/tools/push_cores.htm After sediment recovery, the SW overlying the sediment in the corer was first drawn into a plastic syringe, and the filtered SW through a 0.45-μm filter was then injected into a polypropylene bottle. These SW samples, obtained from a depth of 0 cm below the seafloor (bsf), were refrigerated at 4°C until analysis. After the overlying SW was removed from the corer, the sediment in the corer was subsampled at 5 cm intervals onboard using plastic syringes. Then, the pore water was extracted from the subsamples with a large clamp squeezer during cruise YK06-03 Leg 2 (Manheim [1968]) and by centrifugation during cruise YK08-04 (Bufflap and Allen [1995]).

Analytical Methods

Subsamples of pore fluid and SW to be used for analysis of the concentrations and carbon isotopic compositions of dissolved CH_4, C_2H_6, and total carbon dioxide ($CO_2 = H_2CO_3 + HCO_3^- + CO_3^{2-}$) were transferred to 2 cm^3 glass vials containing H_3NSO_3 to convert the total dissolved carbonate to CO_2 gas and $HgCl_2$ to stop microbial activity. Subsamples of pore fluid and SW to be used for analysis of dissolved chemical components other than the aforementioned dissolved gases were transferred into 4 cm^3 polypropylene bottles. The fluid samples in the polypropylene bottles were measured for NH_4 (= $[NH_4^+] + [NH_3]$) and Si (= $[H_4SiO_4] + [H_3SiO_4^-]$) concentrations in the shipboard laboratory onboard the tender (Gieskes et al. [1991]).

After the fluid samples were transported to the laboratory on land, Cl$^-$ concentrations were measured by Mohr titration, and SO_4^{2-} concentrations were measured by ion chromatography (Tsunogai and Wakita [1995]). The K concentrations in the fluid samples were analyzed by Zeeman-type atomic adsorption spectrometry, and concentrations of the other major and minor elements, Na, Mg, Ca, B (= $[B(OH)_3] + [B(OH)_4^-]$), Sr, Li, and Ba, were analyzed by inductively coupled plasma-atomic emission spectrometry (ICP-AES) (Murray et al. [2000]). The concentrations and carbon isotopic compositions of CH_4, C_2H_6, and CO_2 in the samples for dissolved gas analysis (in the 2 cm^3 vials) were measured by isotopic ratio mass spectrometry (Tsunogai et al. [2002]; Miyajima et al. [1995]). The analytical precision of each measurement technique is given in Table 2. In these measurements on land, the weights of the samples for all analyses were measured, and the concentrations are represented in units per kilogram.

Table 2: Analytical methods and errors for the measurement of chemical components in the pore water

Component	Analytical method	Analytical error
pH (25°C, 1 atm)	Potentiometry	0.2%
Alkalinity	Potentiometric titration	1.2%
NH4+	Colorimetry	7.5%
Si	Colorimetry	1%
Cl-	Titration	1%
SO42-	Ion chromatography	4%
K	Atomic absorption spectrometry	3%
Na	ICP-AES	7%

Ca	ICP-AES	4%
Mg	ICP-AES	1.2%
B	ICP-AES	3%
Sr	ICP-AES	3.5%
Li	ICP-AES	6%
Ba	ICP-AES	10%
δ13CCH4	Mass spectrometry	0.3‰
δ13CC2H6	Mass spectrometry	0.3‰
δ13CΣCO2	Mass spectrometry	0.3‰
δ18OH2O	Mass spectrometry	0.1‰
δDH2O	Mass spectrometry	1‰

ICP-AES, inductivity coupled plasma-atomic emission spectrometry.

Toki et al.

Toki et al. Earth, Planets and Space 2014 66:137, doi:10.1186/s40623-014-0137-3

The O isotopic composition of the water of the fluid samples was analyzed using an equilibration method with $NaHCO_3$ as the reagent (Ijiri et al. [2003]), and the H isotopic composition was analyzed by the Cr reduction method (Itai and Kusakabe [2004]). The isotopic compositions are represented by notation relative to standard materials: Vienna Pee Dee Belemnite (VPDB) for carbon isotopes and Vienna Standard Mean Ocean Water (VSMOW) for O and H isotopes.

$$\delta^{13}C_{carbon} = \left(\left({}^{13}C_{carbon} / {}^{12}C_{carbon} \right)_{sample} \right.$$

$$\left. / \left({}^{13}C_{carbon} / {}^{12}C_{carbon} \right)_{VPDB} \right) - 1 \ (‰ \ VPDB)$$

Carbon : CH_4, C_2H_6, ΣCO_2

$$\delta^{18}O = \left(\left({}^{18}O / {}^{16}O \right)_{sample} / \left({}^{18}O / {}^{16}O \right)_{VSMOW} \right)$$

$$- 1 \ (‰ \ VSMOW)$$

$$\delta D = \left((D / H)_{sample} / (D / H)_{VSMOW} \right) - 1 \ (‰ \ VSMOW)$$

RESULTS

Cl⁻, Na, Mg, SO_4^{2-}, K, Ca, B, Si, Sr, Li, NH_4^+, and Ba concentrations in the fluid samples collected during cruises YK06-03 (Leg 2) and YK08-04 are listed in Table 3, together with the concentration ratios of dissolved CH_4 and C_2H_6 (CH_4/C_2H_6), the C isotopic compositions of dissolved CH_4, C_2H_6, and ΣCO_2 ($\delta^{13}C_{CH4}$, $\delta^{13}C_{C2H6}$, and $\delta^{13}C_{\Sigma CO2}$) and the O and H isotopic compositions of the water ($\delta^{18}O_{H2O}$ and δD_{H2O}). Vertical profiles of the concentrations of Cl⁻, SO_4^{2-}, B, Li, and NH_4^+ and the isotopic information in the pore fluids are shown in Figure 2, since they are a focus of this paper. In this paper, we refer to sampling points inside the bacterial mats and tube worm colonies as 'cold seep sites' and those outside the bacterial mats as 'reference sites'.

Table 3: Chemical and isotopic compositions of pore water and bottoms SW

Table 3 Chemical and isotopic compositions of pore water and bottoms SW

Dive	Sample	Number	Depth cm bsf	Cl^- mmol/kg	Na mmol/kg	Mg mmol/kg	SO_4^{2-} mmol/kg	K mmol/kg	Ca mmol/kg	B μmol/kg	Si μmol/kg	Sr μmol/kg	Li μmol/kg	NH_4^+ μmol/kg	Ba μmol/kg	CH_4/C_2H_6	$\delta^{13}C_{CO2}$ ‰ VPDB	$\delta^{13}C_{CH4}$ ‰ VPDB	$\delta^{13}C_{C2H6}$ ‰ VPDB	$\delta^{18}O_{H2O}$ ‰ VSMOW	δD_{H2O} ‰ VSMOW
949	C1	0	0	543	470	53.6	28.8	10.3	10.4	422	139	92.9	25.5	5			-31.4	-73.6		+0.16	+1.1
		2	5	533	463	49.0	21.5	11.0	7.0	666	303	81.7	26.0	91			-30.7	-74.0		-0.21	-4.9
		3	10	528	466	48.6	21.5	11.7	6.9	710	316	82.8	29.6	91			-29.4	-78.5		-0.40	-3.5
		4	14		474	50.6	23.6	11.7	8.0	655	287	85.8	27.5	68			-28.6	-77.4		-0.40	-3.4
		5	19	541	462	48.8	32.2	11.4	7.3	659	270	84.2	29.0	75			-28.3	-78.6		+0.32	-4.6
		6	23		497	53.4	23.7	12.0	7.9	659	231	91.0	27.4	59						-0.03	-2.3
	C3	0	0	534	481	54.2	28.1	10.7	10.6	438	164	95.0	26.2	11	0.3		-34.1	-94.5	-52.1		
		1	5	515		42.3	10.9	10.3	5.8	951	585	73.7	23.2	141	0.7	9,490	-36.4	-93.3	-47.5		
		2	10	505	436	39.1	4.6	9.7	2.9	1,193	431	67.4	23.2	153	1.1	12,300	-36.8	-96.3	-47.4		
		3	14	498	432	38.6	3.6	9.8	2.7	1,218	487	67.7	24.0	135	2.9	12,400	-37.2	-87.0	-47.3		
		4	19	503			4.4				378		134		1.9	6,510		-88.2			
		5	23	496	452	39.2	3.0	10.1	2.8	1,283	349	67.5	24.8	137	1.4		-37.2	-88.2			
1062	C1	1	6	531	467	46.3	14.7	11.8	5.4	961		73.7	28.1	188	2.3		-36.6	-86.8		-0.29	
		2	11	517	451	42.5	7.1	11.2	3.4	1,371		65.8	24.9	280	3.8		-36.6	-86.8		-0.29	
		3	16	510	454	41.8	5.2	11.6	2.2	1,213		60.7	23.9	294	5.5		-39.6	-80.8		-0.30	
		4	21	517	441	40.8	3.2	11.2	1.0	1,242		53.9	22.9	297	6.0		-35.4	-92.4		-0.28	
		5	26	503	456	41.4	0.3	10.1	1.0	1,163		54.1	23.0	294	18.8	4,040	-38.9	-86.4	-41.5	-0.40	
	C2	0	0	551	472	52.4	27.5	10.0	10.1	403		89.5	25.3	1			-1.2	-65.0		-0.32	
		1	3	545	473	51.7	28.3	10.6	10.1	424		88.2	27.8	9			-2.7	-75.0		-0.30	
		3	13	540	472	51.6	27.1	10.8	9.9	430		87.4	26.1	27			-4.2	-70.3		-0.21	
		4	18	534	473	51.4	27.2	11.1	9.8	428		87.1	26.8	27			-4.0	-80.4		-0.36	
		5	23	538	471	51.3	27.6	10.9	9.8	437		86.8	26.3	33						-0.32	
	C3	0	0	557	468	52.6	26.5	9.9	10.2	422		91.8	26.1	17			-26.6	-82.0		+0.09	
		1	4	538	466	49.1	22.7	10.4	8.3	468		88.1	26.4	164			-32.2	-82.2		-0.21	
		2	9	544	456	47.2	19.6	10.5	6.6	525		83.0	27.2	144						-0.37	
	C4	4	19	531	456	45.6	16.0	10.5	5.6	637		82.6	26.3	241			-34.1	-80.5		-0.35	
	C4	0	0	548	470	53.2	26.2	9.8	10.4	407		91.5	26.3	2						-0.16	
		1	2	543	468	50.5	27.0	11.8	9.9	469		89.4	28.8	10			-6.6	-58.4		-0.22	

2	7	53.8	47.0	50.4	26.6	11.7	9.8	499		89.3	28.9	12	-11.0	-80.4	-0.18
3	12	53.8	46.9	49.6	25.8	12.0	9.5	502		88.8	23.8	22	-15.7	-81.7	-0.18
4	17	53.6	46.6	48.7	23.7	11.3	9.2	510		86.8	23.3	38	-22.0	-82.5	-0.05
5	22	53.6	46.3	46.7	22.2	12.4	8.7	608		84.4	27.1	56	-25.0	-82.2	-0.04
CS 0	0	54.8	46.7	52.7	26.7	9.9	10.2	402	668	91.3	23.8	3			-0.20
1	4	53.6	46.0	46.0	46.0	19.7	12.7	87		86.2	28.3	97	-29.8	-73.0	-0.33
2	9	53.2	49.1	45.2	18.2	12.4	6.3	721		75.4	28.8	118	-32.0	-73.2	-0.24
3	14	52.8	45.0	40.6	13.3	13.1	4.8	828		69.2	27.2	169	-35.8	-72.9	-0.37
4	19	53.0	45.2	41.9	13.6	11.7	4.8	789		70.3	28.2	131	-35.4	-73.4	-0.41
5	24	54.4	47.4	44.0	15.5	11.8	5.7	800		81.4	26.8	129	-35.3	-72.9	-0.31

Toki et al.

Toki et al. Earth, Planets and Space 2014 66:137, doi: 10.1186/s40623-014-0137-3

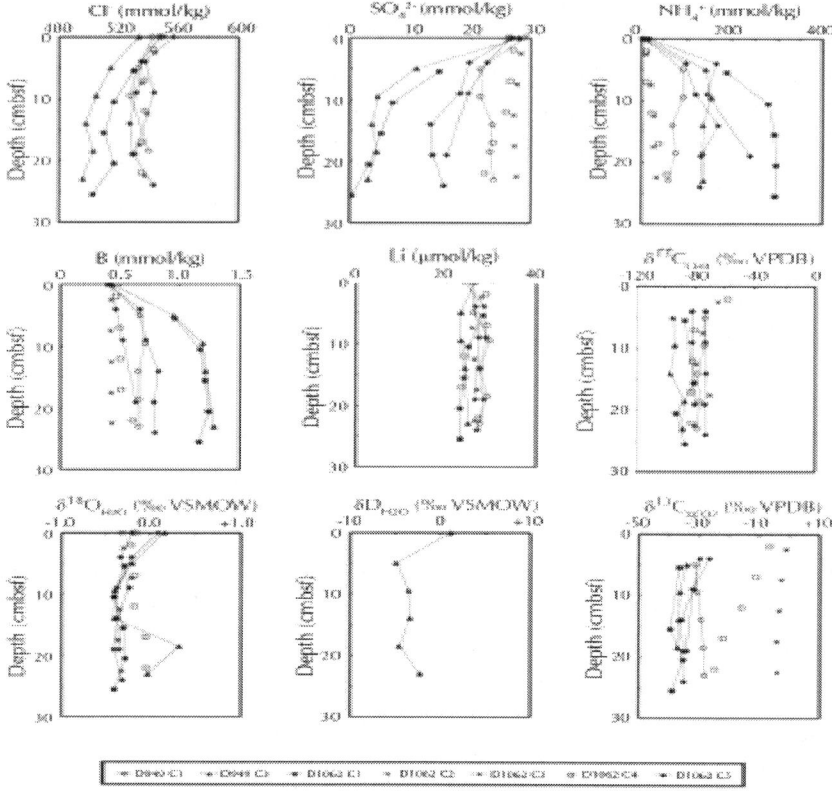

Figure 2: Vertical profiles of chemical and isotope components of pore water from sediments in Oomine Ridge. The points representing samples from cold seep sites are connected by lines to differentiate them from reference site data.

The chemical and isotopic compositions of the fluid samples at cold seep sites differed from those at the reference sites (Figure 2). Regarding the curvatures in the graphs for Cl^- and SO_4^{2-}, the upward curvatures appear to imply upward movement of waters from depth. The occurrence of bacterial mats suggests that some CH_4 escapes to the surface to feed these mats (Gieskes et al. [2005]). The curvatures of $\delta^{18}O_{H2O}$ and δD_{H2O} are also apparent, especially in D949 C3, with some 'flyers' in the δD_{H2O} in the deeper part of C3. In a subsequent section, it is suggested that water flow does occur from greater depths. Therefore, this curvature must occur, with SW mixing occurring in the upper 10 cm of the cores.

The Cl⁻ concentrations at the seafloor (depth = 0 cm bsf) were averaged to be 547 mM with a standard deviation of 8 mM (Table 4). These samples correspond to bottom SW. The Cl⁻concentrations in these samples were almost consistent with that of North Pacific deep SW (548 mM; Reid [2009]), suggesting the accuracy of the reported values. However, the standard deviation, at 1.5%, is larger than the 1% analytical precision of the Mohr method (Table 1). This deviation is due not only to an influence of low-Cl⁻ seep fluids because otherwise, the values would be lower than that of North Pacific deep SW. However, we also detected higher values than that of North Pacific deep SW (Table 3). The chemical characterization of Cl⁻ is generally nonreactive, in which Cl⁻ increases only by dissolution of evaporites. Evaporites rarely exist in natural environments and occur only around dry regions. Moreover, the existence of evaporites has not been reported near Nankai Trough. Unfortunately, we did not determine the reason for the higher values than that of North Pacific deep SW, although the lower Cl⁻ concentration in the overlying SW at D949 C3 where the steepest curvature was observed in vertical profiles of chemical and isotopic compositions may be due to the influence of low-Cl⁻ seep fluids (Figure 2).

Table 4: Chemicaland isotopic compositions of end members for the estimation in this study

Sources	Cl⁻	B mmol/kg	δ18OH2O‰ VSMOW	δDH2O‰ VSMOW	Reference
SW	547 ± 8	0.4116 ± 0.015	−0.09 ± 0.21	+1.1 ± 1.0	This study
CMD	0	23 ± 8	ND	ND	Toki et al. [2013]
SPF	549 ± 4	0.217 ± 0.057	−2.5 ± 0.4	−10.0 ± 2.8	Expedition 315 Scientists [2009a]
MH	0	0	+0.6 ± 0.4	+8.0 ± 2.8	Expedition 315 Scientists [2009a]; Maekawa [2004]
Oomine	498	1.3	−0.4	−3.4	This study

SW, seawater; CMD, clay mineral dehydration; SPF, shallow pore fluid; MH, methane hydrate; Oomine, D949 C3-3; ND, not determined.

Toki et al.

Toki et al. Earth, Planets and Space 2014 66:137, doi:10.1186/s40623-014-0137-3

DISCUSSION

Origin of B in of Pore Fluids at Cold Seep Sites on the Oomine Ridge

At cold seep sites on the Oomine Ridge, the B concentration in the pore fluids increased with depth (Figure 2). Possible sources of B are organic matter desorption, which occurs at relatively low temperatures (You et al. [1993b]; Brumsack et al. [1992]), and smectite-illite alteration, which occurs between 50°C and 160°C (You et al. [1996]). Organic matter desorption is related to organic matter decomposition and thus results in well-correlated B and NH_4^+ concentrations with $\Delta B/\Delta NH_4^+$ ratios of 0.1 mol/mol (Teichert et al. [2005]). In this study, $\Delta B/\Delta NH_4^+$ was about 4 mol/mol at the cold seep sites on the Oomine Ridge; thus, the ratio demonstrated B enrichment by a factor of 40 compared with the expected ratio for organic matter desorption (Figure 3). This excess B suggests that B derived from smectite-illite alteration occurring in a higher-temperature environment is supplied to the pore fluids in the surface sediments at these cold seep sites and that the supplied fluids were subjected to temperatures between 50°C and 160°C.

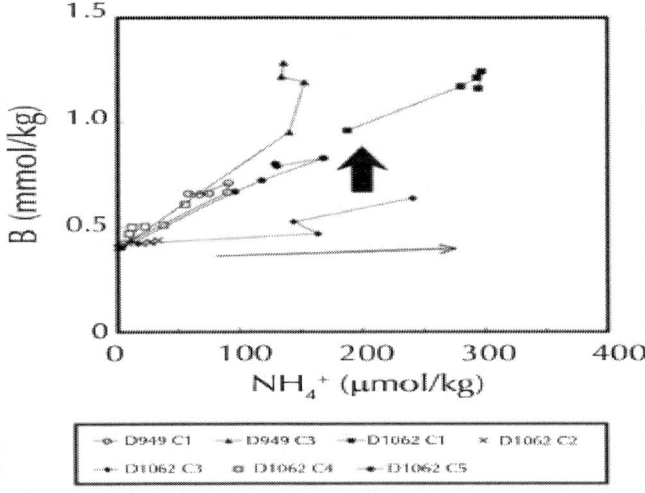

Figure 3: Relationship between NH4+and B concentrations in pore water from the Oomine Ridge. The thin arrow represents B/ NH$_4^+$ratios of approximately 0.1 mol/mol for B derived from only organic matter desorption. The thick arrow indicates the manner in which B concentrations are enriched compared with those derived from only organic matter desorption. The sample numbers are given in Table 1.

At Site C0001, near the surface trace of a megasplay fault in analogy with the Oomine Ridge, the coring was operated up to 458 m bsf during Integrated Ocean Drilling Program (IODP) Expedition 315 (Expedition 315 Scientists [2009a]). The results indicated that the heat flow was 47 mW/m^{2}and that there was no B enrichment. On the contrary, at Site C0002 on the northern rim of Kumano Basin (Expedition 315 Scientists [2009b]), the maximum depth reached during the Expedition 315 drilling was 1,057 m bsf. In addition, the heat flow was 56 mW/m^2 (Harris et al.[2011]) and there was no B enrichment. These heat flow data indicate that the high-temperature zone in which clay mineral dehydration (CMD) can occur was not reached by drilling in Kumano Basin. Thus, the lack of B enrichment at both sites implies that B-rich fluids at either depth did not pass through the hanging wall of the megasplay (C0001) or Kumano Basin (C0002). This result leads to the question of how B-rich fluids can flow to the Oomine Ridge.

In Nankai Trough off Muroto, high B concentrations in pore fluids up to 3 mmol/kg are shown in the décollement zone, which is attributed to fluid flow in the décollement zone (You et al. [1993a]). In the Japan Trench forearc, B shows an increase in pore fluids, which is also attributed to fluid advection in sediments (Deyhle and Kopf [2002]). In all cases, lower chlorides than that of SW were noted. Thus, inputs of B by fluid flow seem to be clear in these areas. In the Nankai forearc off Kumano, fluids may flow through the megasplay fault, although we were unable to detect corresponding B enrichment at Site C0004 penetrating the megasplay fault (Expedition 316 Scientists [2009a]). Very few cold seep sites have been observed near the surface trace of the megasplay fault near Site C0004 (Ashi et al. [2009b]). These observations suggest that the megasplay near Site C0004 is not an active pathway for reductive fluids. A possible explanation for the different character of the megasplay between Oomine and C0004 is the difference in tectonic and hydrologic activities along the strike of the megasplay fault (e.g., Kimura et al. [2011]).

Isotopic Compositions of Pore Fluids at Cold Seep Sites on the Oomine Ridge

At the cold seep sites on the Oomine Ridge, the observed $\delta^{18}O_{H2O}$ and δD_{H2O} were negative (Figure 2), which led Toki et al. ([2004]) to conclude that these fluids originated from groundwater. As suggested in 'Origin of B in of pore fluids at cold seep sites on the Oomine Ridge' section, however, if the fluids were derived from CMD, the values of $\delta^{18}O_{H2O}$ and δD_{H2O} would be positive and negative, respectively (Magaritz and Gat [1981]). It is possible that during their ascent to the seafloor, fluids derived from CMD mixed with fluids in sediments shallower than the estimated depth range of 1.5 to 3.5 km bsf, where temperatures from 50°C to 160°C enable CMD to occur. Such chemical and isotopic features have been reported in the Barbados subduction zone, although the distribution of δD_{H2O} has not been explained by any possible processes in the sediments (Vrolijk et al. [1990], [1991]).

The chemical and isotopic compositions of the Oomine Ridge cold seep pore fluids also reflect the mixing of low Cl⁻, $\delta^{18}O$, and δD fluids with SW (Figure 2). The data for Cl⁻ and $\delta^{18}O$, however, are scattered (Figure 2), implying that the pore fluids did not result from the simple mixing of two sources such as SW and another end member. Among the cold seep samples, sample D949 C3-3 ($Cl^- = 498$ mmol/kg, $B = 1.3$ mmol/kg, $\delta^{18}O_{H2O} = -0.4‰$, $\delta D_{H2O} = -3.4‰$) differs most from SW with respect to these values (Figure 2). If the data of D949 C3-3 can be explained by some end members, the other data also can be explained by those end members with different mixing ratios. Therefore, we examined various possible sources in which the mixing with freshwater derived from CMD might explain the chemical and isotopic compositions of the pore fluids in the cold seeps on the Oomine Ridge.

Possible Sources of Fluids in Cold Seeps on the Oomine Ridge

Seawater

First, we calculated the chemical and isotopic compositions of the SW samples from D949 (C1-0 and C3-0) and D1062 (C2-0, C3-0, C4-0, and C5-0), which consisted of SW overlying the sediment in each corer collected just before the sediments were sampled (Table 3). The chemical and isotopic compositions in these samples were averaged as $Cl^- = 547 \pm 8$ mmol/kg, $B = 0.416 \pm 0.015$ mmol/kg, $\delta^{18}O_{H2O} = -0.09 \pm 0.21‰$, and $\delta D_{H2O} = +1.1 \pm 1.0‰$. All of these values fall within the range of those of North Pacific deep SW (Reid [2009]); therefore, we adopted these values for the chemical and isotopic compositions of SW in this study.

Freshwater Derived from CMD

The chemical and isotopic compositions of freshwater derived from CMD ($\delta^{18}O_{CMD}$ and δD_{CMD}) depend on those of the clay minerals ($\delta^{18}O_{clay}$ and δD_{clay}) and the equilibrium temperature T (K) of the reaction. The isotopic fractionation between clay minerals and ambient pore fluids for $\delta^{18}O$ (Sheppard and Gilg [1996]) and δD (Capuano [1992]) is expressed as a function of the reaction temperature:

$$\delta^{18}O_{clay} - \delta^{18}O_{CMD} = \frac{2.55 \times 10^6}{T^2} - 4.05$$

(1)

$$\delta D_{clay} - \delta D_{CMD} = -\frac{4.53 \times 10^4}{T} + 94.7$$

(2)

In these equations, we considered $\delta^{18}O_{clay}$ to range from +17‰ to +26‰ and δD_{clay} to range from −95‰ to +33‰ because these are the reported value ranges for clay minerals in marine sediments (Savin and Epstein [1970]; Yeh [1980]; Capuano [1992]). Then, using Equations 1 and 2, we calculated $\delta^{18}O_{CMD}$ to range between −3‰ and +17‰ and δD_{CMD} to range between −85‰ and +79‰ for a temperature range of 50°C to 160°C. The B concentration of freshwater derived from CMD is unknown; therefore, we used the value of the Kumano mud volcanoes near the study area (23 ± 8 mmol/kg) as the reference value (Toki et al. [2013]). Since the value of 23 mmol/kg is taken from the mud volcano site, the accuracy should be lower, and the error would be larger than 8 mmol/kg. But this value is essential to the modeling, and it should be stated for the fairness.

Shallow Pore Fluids in Nankai Accretionary Prism Sediment

From 2007 to 2008, during IODP Expeditions 315 and 316, D/V *Chikyu* drilled into the slope of the Nankai accretionary prism to a depth of up to 1,052 m bsf and recovered pore fluids from the sediments (Figure 1b) (Ashi et al. [2009a]; Screaton et al. [2009]). Pore fluid $\delta^{18}O_{H2O}$ and δD_{H2O} values at Sites C0001 and C0002, drilled during Expedition 315, have been reported (Expedition 315 Scientists [2009b]), although at Sites C0004 and C0008, drilled during Expedition 316, only pore fluid $\delta^{18}O_{H2O}$ has been reported (Expedition 316 Scientists [2009a], [b]). The $\delta^{18}O_{H2O}$ and δD_{H2O} at all four sites, however, have been measured by Dr. H. Tomaru using the method described by Expedition 315 Scientists ([2009a]). Dr. Tomaru distributed the data to all researchers associated with the drilling expeditions, including onshore researchers requesting the data. We therefore used these data to construct vertical profiles of $\delta^{18}O_{H2O}$ and δD_{H2O} (Figure 4). At depths deeper than 200 m bsf at all sites drilled during Expeditions 315 and 316, $\delta^{18}O_{H2O}$ and δD_{H2O} were nearly constant: $\delta^{18}O_{H2O}$ varied between −4.5‰ and −2‰, and δD_{H2O} varied between −15‰ and −10‰ (Figure 4). On the contrary, at depths shallower than 200 m bsf at all sites, these values gradually became close to that of SW. These gradients suggest that a fluid source with constant $\delta^{18}O_{H2O}$ and δD_{H2O} below 200 m bsf would mix with SW above 200 m bsf. In such cases, the fluid source below 200 m bsf would be recognized as an end member in the shallow sediments of the Nankai accretionary prism slope (below 200 m bsf and above 1.5 to 3.5 km bsf). In this study, we refer to the fluids in the sediments of the Nankai accretionary prism slope below 200 m bsf and above 1.5 to 3.5 km bsf as shallow pore fluids (SPF). The SPF ubiquitous in the sediments of the Nankai accretionary prism slope would be mixed with deep-sourced fluids derived from CMD before seeping at the cold seep sites on the Oomine Ridge. In our subsequent discussion, we use the following chemical and isotopic compositions of SPF from Site C0001 on the outer ridge of the Nankai accretionary prism as SPF values: $Cl^- = 549 \pm 4$ mmol/kg, $B = 0.217 \pm 0.057$ mmol/kg, $\delta^{18}O = -2.5 \pm 0.4$‰, and $\delta D = -10.0 \pm 2.8$‰ (Expedition 315 Scientists[2009a]).

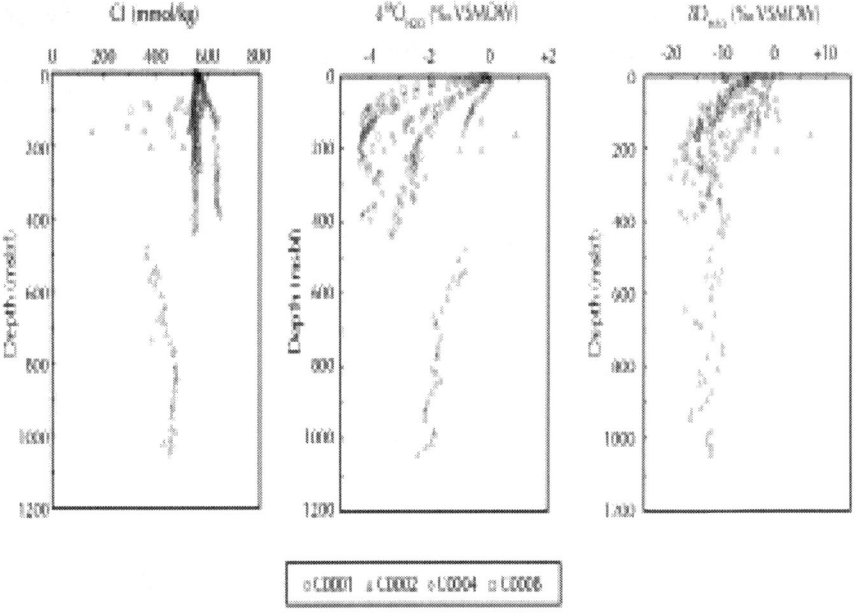

Figure 4: Vertical profiles of Cl-, δ18OH2O, and δDH2O. Vertical profiles of Cl-, $\delta^{18}O_{H2O}$, and δD_{H2O} in the pore fluids at Sites C0001, C0002, C0004, and C0008 in the Nankai accretionary prism.

Freshwater from Methane Hydrate Dissociation

A final factor that can influence the chemical and isotopic compositions of the pore fluids is the freshwater derived from MH dissociation. MH has never actually been recovered from Site C0001, which is situated in a position similar to the Oomine Ridge (Ashi et al. [2009a]). However, discontinuous bottom-simulating reflectors (BSRs) suggesting the presence of MH have been observed beneath the slope of the Nankai accretionary prism (e.g., Colwell et al. [2004]). Moreover, MH was recovered from several hundred meters below the seafloor at Site C0008, which is near the surface trace of another megasplay fault on the seaward side of the Oomine Ridge (Screaton et al. [2009]). In general, MH is recovered where sand layers occur (Ginsburg et al. [2000]), although beneath the Nankai accretionary prism slope, which is composed mainly of silty clay, dispersed MH may be present (Screaton et al. [2009]). Taken together, these findings suggest that it is possible for freshwater derived from MH dissociation to contribute to the fluids supplied to the cold seeps on the Oomine Ridge. Therefore, we also considered freshwater from MH dissociation in our estimation of possible contributions to the seepage fluid.

When MH forms in sediments, it consists mainly of CH_4 and water, excluding salt from the ambient SW (e.g., Sloan and Koh [2008]). When MH is recovered during drilling, the MH dissociates, depending on the temperature and pressure conditions, to release CH_4 and water (Hesse and Harrison [1981]; Ussler and Paull [1995]). In vertical profiles of pore fluids, samples influenced by MH dissociation are characterized by a negative Cl^- concentration spike and positive $\delta^{18}O_{H2O}$ and δD_{H2O} values (Kvenvolden and Kastner [1990]). Experimentally determined $\delta^{18}O_{H2O}$ and δD_{H2O} fractionation factors during MH formation show a shift to heavier values from ambient water to the formation water; that is, $\Delta\delta^{18}O$ shifts from +2.8‰ to +3.2‰ and $\Delta\delta D$ shifts from +16‰ to +20‰ (Maekawa [2004]; Maekawa and Imai [2000]). Here, we adopt as the reference value $\Delta\delta^{18}O = +3.1‰$, which was obtained in the Nankai Trough gas hydrate area (Tomaru et al.[2004]).

If MH has formed in the vicinity of the outer ridge, then, given the composition of the SPF at Site C0001 ($Cl^- = 549 \pm 4$ mmol/kg, $\delta^{18}O_{H2O} = -2.5 \pm 0.4‰$, and $\delta D_{H2O} = -10.0 \pm 2.8‰$) and the experimentally determined isotope fractionation values of $\Delta\delta^{18}O = +3.1‰$ and $\Delta\delta D = +16‰$ to +20‰, the isotopic composition of the formation water of MH can be estimated to be $\delta^{18}O_{H2O} = +0.6 \pm 0.4‰$ and $\delta D_{H2O} = +6.0 \pm 2.8‰$ to +10.0±2.8‰. We used these values to calculate the contribution of MH dissociation to the pore fluids on the Oomine Ridge.

Mixing Model for the Formation of Fluids Supplied to the Oomine Ridge Cold Seeps

We used a mixing model to explain the compositions of pore fluid at the Oomine Ridge cold seeps. Using the end-member values listed in Table 4, we solved the following equations:

$$Cl^-_{Oomine} = X \times Cl^-_{SW} + Y \times Cl^-_{CMD} + Z \times Cl^-_{SPF} + W \times Cl^-_{MH}, \tag{3}$$

$$B_{Oomine} = X \times B_{SW} + Y \times B_{CMD} + Z \times B_{SPF} + W \times B_{MH}, \tag{4}$$

$$\delta^{18}O_{Oomine} = X \times \delta^{18}O_{SW} + Y \times \delta^{18}O_{CMD} + Z \times \delta^{18}O_{SPF} + W \times \delta^{18}O_{MH}, \tag{5}$$

$$\delta D_{Oomine} = X \times \delta D_{SW} + Y \times \delta D_{CMD} + Z \times \delta D_{SPF} + W \times \delta D_{MH}, \tag{6}$$

$$X + Y + Z + W = 1, \tag{7}$$

where each source is denoted by a subscript previously defined. X, Y, Z, and W denote the mixing ratios of SW, CMD, SPF, and MH dissociation, respectively. The calculation outline is schematically drawn in Figure 5. Assuming the feasible combination of $\delta^{18}O_{CMD}$ and δD_{CMD} given by Equations 1 and 2 for 50°C to 160°C as the reaction temperature of CMD, we calculated the mixing ratios where the Oomine seep values lie on the same plane as that of the combination of SW, CMD, SPF, and MH values. For obtained mixing ratios, we verified the existence of a solution in which the predicted B concentration coincides with the observed B concentration; thus, we can accept the results. At first, we considered three-end-member mixing model. One of four mixing ratios (X-W) is forced to zero, and all Equations 3, 4, 5, 6, and 7 are used to statistically obtain the other three ratios. Although several attempts were made by normalizing the coefficients to the same magnitude, all four cases resulted in prediction error much larger than 10%. Thus, neglecting any factor of the four ratios was rejected.

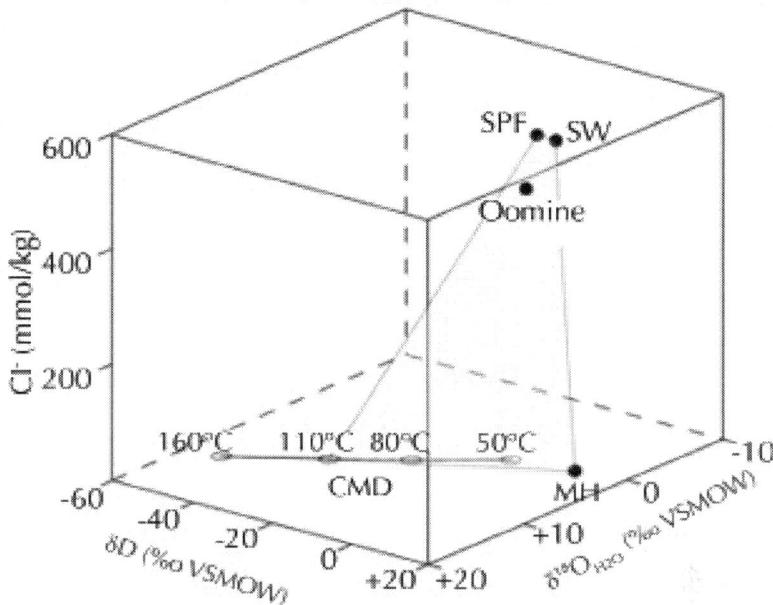

Figure 5: Relationship among Cl-,δ18OH2O, and δDH2Oof sources for the pore fluids. Relationship among Cl⁻, $\delta^{18}O_{H2O}$, and δD_{H2O} of sources for the pore fluids at cold seep sites in the Oomine Ridge. The plot of seawater (SW) is Cl⁻ = 547 mmol/kg, $\delta^{18}O_{H2O}$ = −0.09‰, and δD_{H2O} = +1.1‰. The plot of shallow pore fluids (SPF) is Cl⁻ = 549 mmol/kg, $\delta^{18}O_{H2O}$ = −2.5‰, and δD_{H2O} = −10.0‰. The plot of the pore fluids at cold seep sites in the Oomine Ridge (Oomine) is represented by D949 C3-3, Cl⁻ = 498 mmol/kg, $\delta^{18}O_{H2O}$ = −0.40‰, and δD_{H2O} = −3.4‰. The plots of the clay-derived freshwater (CMD) are on the theoretical curve, drawn by the calculation of theoretical $\delta^{18}O_{H2O}$ and δD_{H2O} values of water at 50°C to 160°C assuming equilibrium fractionation between pore water and clay minerals according to Sheppard and Gilg ([1996]) using clay minerals $\delta^{18}O_{clay}$ = +21.5‰, a medium value for an example within a reported range of +17‰ to +26‰ for $\delta^{18}O_{clay}$, and that reported by Capuano ([1992]) using clay minerals δD_{clay}, and a medium value for an example within a reported range of −50‰ to +43‰ for δD_{clay}. In addition, the plot of freshwater derived from methane hydrate (MH) is Cl⁻ = 0 mmol/kg, $\delta^{18}O_{H2O}$ = +0.3 to +0.7‰, and δD_{H2O} = +6.0 to +10.0‰. We determined the mixing

ratios for each sources, as the Oomine plot is in one plane with SW, SPF, CMD, and MH, represented by the shaded quadrangle.

When we consider a mixing model with four components (end members) including SW, freshwater derived from CMD, SPF, and freshwater derived from MH dissociation, several solutions are possible. For example, for an equilibrium temperature of 160°C, we obtained the solution $(X, Y, Z, W) = (0.47, 0.044, 0.44, 0.047)$ and $(\delta^{18}O_{CMD}, \delta D_{CMD}) = (+16.5‰, +3‰)$; for an equilibrium temperature of 110°C, we obtained $(X, Y, Z, W) = (0.70, 0.042, 0.21, 0.048)$ and $(\delta^{18}O_{CMD}, \delta D_{CMD}) = (+3.7‰, -59‰)$. Using other clay mineral compositions and equilibrium temperatures, other solutions are possible. In particular, for $(\delta^{18}O_{clay}, \delta D_{clay}) = (+26‰, -95‰)$ and for an equilibrium temperature of 50°C, we obtained $(X, Y, Z, W) = (0.67, 0.042, 0.24, 0.048)$ and $(\delta^{18}O_{CMD}, \delta D_{CMD}) = (+5.6‰, -50‰)$. These results indicate that although it is not possible with this model to constrain the equilibrium temperature, the four-end-member model can nevertheless explain the seepage fluid compositions. We conclude that both MH dissociation and SPF contribute to the pore fluids in the cold seeps on the Oomine Ridge. The results indicate that the contributions of SW and SPF, SW > 40% and SPF = 20% to 50%, are dominant, followed by freshwater from clay minerals, CMD = approximately 4%, and MH dissociation, MH = approximately 5%, which contributes the least. The most important finding is that pore fluids at cold seep sites on the Oomine Ridge cannot be explained without considering CMD, SPF, and MH dissociation.

Behavior of Pore Fluids in Sediments off Kumano

At the seepage sites on the Oomine Ridge, the observed B concentrations were greater than could be explained by organic matter degradation, suggesting that the seeps at the site are supplied with fluids derived from CMD at 50°C to 160°C Toki et al. ([2013]) have shown that Li as well as B is supplied to the Kumano mud volcanoes and that the presence of Li can be attributed to the fluids passing through layers with temperatures of 150°C to 160°C before reaching the seepage sites. The isotopic compositions of the pore fluids of the mud volcanoes also indicate derivation from CMD (Toki et al. [2013]). Thus, the mud volcano fluids ascend from about 4 km bsf (Figure 6; Source A). Moreover, these deep-origin fluids are not overprinted by other fluids in the shallow sediments during their ascent. This scenario can explain the concentrations of other components derived from great depth, including B and Li. The sediment thickness in Kumano Basin is only about 2 km (Expedition 315 Scientists [2009b]); therefore, the rock around Source A at 4 km bsf is likely composed of old accretionary sediments in the lower part of the accretionary prism (Figure 6).

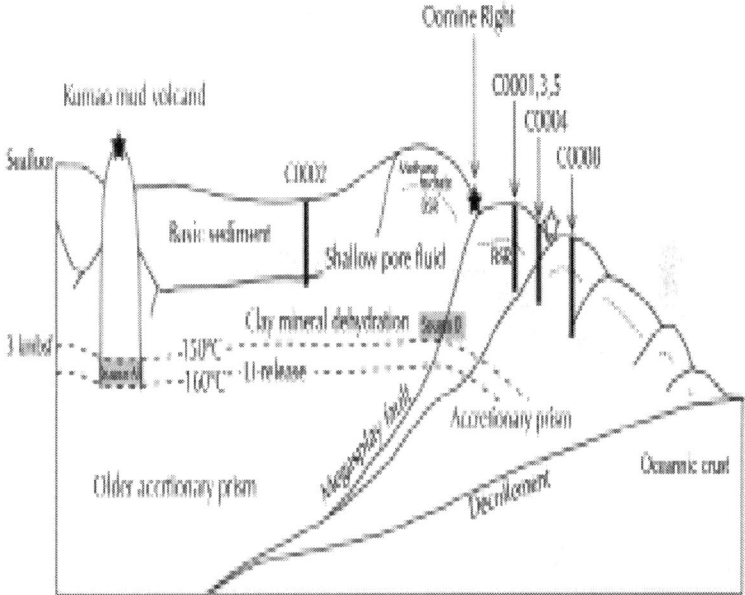

Figure 6: Schematic diagram showing the distribution and migration of pore fluids in sediments off Kumano. The solid stars indicate cold seep sites at the Oomine Ridge; open star indicates cold seep sites at the other area. The black vertical bars indicate IODP sites. The dotted lines are bottom-simulating reflectors (BSRs) that delineate a base of MH, and the dashed lines are temperature contours of 150°C and 160°C. The shaded zone indicates a reservoir of shallow pore fluid (SPF). Source A indicates the source of mud volcano fluids characterized by B and Li enrichment, whereas Source B indicates a source of the Oomine Ridge fluids rich in B but not Li.

The results obtained in this study suggest that the pore fluid composition at the seepage sites on the Oomine Ridge reflects contributions from other sources, in addition to freshwater from CMD. Therefore, as the source fluids ascend from depth below the seafloor, they likely mix with pore fluids in the shallower layers of the accretionary prism before finally seeping out at the seafloor. However, when we considered the contributions only from SPF in the shallow accretionary prism, which is represented by fluids obtained during deep drilling, freshwater derived from CMD, and SW, the resulting contribution ratios could not explain the observed B concentration ('Shallow pore fluids in Nankai accretionary prism sediment' section). When we also considered a contribution of fluids derived from MH dissociation, we obtained the following contribution ratios in the seepage fluid ('Freshwater derived from CMD' section): about 4% fluid derived from CMD, about 5% fluid derived from MH dissociation, 20% to 50% SPF of the accretionary prism, and more than 40% SW (Figure 6; Source B). These ratios can explain the observed chemical and isotopic compositions of the pore fluids at the cold seep sites on the Oomine Ridge. Therefore, we conclude that fluids from MH dissociation, CMD, and SPF likely contribute to the pore fluids of the cold seeps.

The results of our estimation of fluid sources suggest that the mode of transport differs between the Oomine Ridge and the Kumano mud volcanoes. In the case of the Kumano mud volcanoes, both source fluids in sediments and the sediments themselves ascend to the seafloor, whereas on the Oomine Ridge, the source fluids mix with SPF sediments and with fluid from MH dissociation as they ascend to the seafloor, although the sediments themselves do not ascend. These different modes of fluid transport are consistent with hydrocarbon distribution differences between the Kumano mud volcanoes and the Oomine Ridge. Hydrocarbons of thermogenic origin are found only in the Kumano mud volcanoes, even though the fluids supplied to both the seepage sites on the Oomine Ridge and the Kumano mud volcanoes originate in environments with temperatures of more than 50°C (Toki et al. [2013]). CH_4 of microbial origin is distributed ubiquitously in the shallow sediments above several hundred meters below the seafloor in the accretionary prism off Kumano (Toki et al. [2012]). Thus, sediments containing hydrocarbons of thermogenic origin rise to the seafloor within the Kumano mud volcanoes and are observed in the pore fluids in the subsurface sediments. In contrast, the fluids supplied to the Oomine Ridge contain CH_4 of microbial origin from the shallow sediments through which the fluids passed during their ascent to the seafloor.

On the basis of the correlation between decreases in the isotopic compositions of the pore fluids and Cl⁻ concentration, Toki et al. ([2004]) inferred that the source of the seepage fluids on the Oomine Ridge is laterally transported meteoric water. The drilling beneath the seafloor since 2007 has revealed chemical and isotopic compositions of the pore water in the sediments to several hundred meters below the seafloor. The isotopic compositions of the pore water in the shallow sediments of Nankai accretionary prism both had negative values, similar to those of meteoric water (Figure 2). In this study, we focused on B concentrations in the pore fluids, which showed that the seepage fluids on the Oomine Ridge as well as Kumano mud volcano fluids are influenced by freshwater derived from CMD. The source fluids of the cold seeps can thus become mixed with SPF during their ascent to the seafloor. We observed no Li anomaly on the Oomine Ridge; thus, the source fluids did not pass through environments with temperatures above 150°C. We cannot rule out the possibility, however, that the characteristics of the original fluids in the deeper environments have been changed by mixing with SPF in the shallower sediments during the ascent. In the future, by conducting experiments with water and rock to determine trace elements not analyzed by You et al. ([1996]), a tracer that sensitively records information from deeper environments should be identified and utilized.

CONCLUSIONS

The results of this study are summarized in the following points:

- During cruises YK06-03 and YK08-04 of the tender *Yokosuka*, we collected pore water samples at cold seep sites on the Oomine Ridge. In these pore fluid samples, we found B derived from smectite-illite alteration, which suggests that the fluids were derived from environments with temperatures between 50°C and 160°C.

- Our estimation of the source fluids based of a mixing model including a contribution of fluid from MH dissociation indicated that a mixture containing about 4% freshwater derived from CMD, about 5% freshwater derived from MH dissociation, 20%

to 50% SPF from accretionary prism sediments, and more than 40% SW can explain the chemical and isotopic compositions of the cold seep fluids on the Oomine Ridge.

- The fluids seeping on the Oomine Ridge are transported from depth via faults and are mixed with SPF of the accretionary prism sediments and freshwater derived from MH dissociation prior to reaching the seafloor. This transport mode is clearly different from that of the fluids in mud volcanoes, which ascend together with sediments and do not mix with SPF.

AUTHORS' CONTRIBUTIONS

TT interpreted the data and wrote the manuscript. RH carried out chemical analyses, had a discussion and wrote the first draft. AI carried out the oxygen isotope measurements, had a discussion and modified the manuscript. UT thoroughly supported the isotopic analyses and modified the manuscript. JA organized the sampling campaign and modified the manuscript. All authors read and approved the final manuscript.

ACKNOWLEDGEMENTS

The authors thank the *Shinkai6500* operation team and the captain and crew of the tender*Yokosuka* during cruises YK06-03 and YK08-04 for their continued dedication. We are grateful to Profs. S. Ohde, T. Oomori, and T. Matsumoto for their valuable comments that improved an earlier version of the manuscript. We also thank Prof. J. M. Gieskes and an anonymous reviewer for their constructive suggestions, as well as Dr. Masataka Kinoshita as the guest editor. We would like to express our sincere gratitude to Dr. Hitoshi Tomaru for providing the $^{18}O_{H2O}$ and D_{H2O} data. This research was supported by a Grant-in-Aid for Scientific Research on Innovative Areas KANAME project. Moreover, during the writing of this paper, the authors were supported by the International Research Hub Project for Climate Change and Coral Reef/Island Dynamics from the University of the Ryukyus.

REFERENCES

1. Aharon P, Fu B (2000) Microbial sulfate reduction rates and sulfur and oxygen isotope fractionations at oil and gas seeps in deepwater Gulf of Mexico. Geochim Cosmochim Acta 64:233-246

2. Aloisi G, Drews M, Wallmann K, Bohrmann G (2004) Fluid expulsion from the Dvurechenskii mud volcano (Black Sea): Part I. Fluid sources and relevance to Li, B, Sr, I and dissolved inorganic nitrogen cycles. Earth Planet Sci Lett 225:347-363

3. Ashi J, Lallemant S, Masago H (2009a) Proceedings of the Integrated Ocean Drilling Program, 314/315/316.Kinoshita M, Tobin H, Ashi J, Kimura G, Lallemant S, Screaton EJ, Curewitz D, Masago H, Moe KT (eds) Integrated Ocean Drilling Program Management International, Inc, Washington DC.

4. Ashi J, Tsuji T, Ikeda Y, Morita S, Hashimoto Y, Sakaguchi A, Ujiie K, Saito S, Kuramoto S (2009b) Seafloor expressions of fault activities in the Nankai accretionary prism off Kumano. In: AGU Fall Meeting. San Francisco, 14 December 2009

5. Brumsack HJ, Zuleger E, Gohn E, Murray RW (1992) Stable and radiogenic isotopes in pore waters from leg 127, Japan Sea. Proceedings of the Ocean Drilling Program, Scientific Results, 127/128: Ocean Drilling Program. College Station, pp 635–650

6. Bufflap SE, Allen HE (1995) Sediment pore water collection methods for trace metal analysis: A review. Water Res 29:165-177

7. Capuano RM (1992) The temperature dependence of hydrogen isotope fractionation between clay minerals and water: Evidence from a geopressured system. Geochim Cosmochim Acta 56:2547-2554

8. Chao H, You C, Wang B, Chung C, Huang K (2011) Boron isotopic composition of mud volcano fluids: Implications for fluid migration in shallow subduction zones. Earth Planet Sci Lett 305:32-44

9. Colwell F, Matsumoto R, Reed D (2004) A review of the gas hydrates, geology, and biology of the Nankai Trough. Chem Geol 205:391-404

10. Dählmann A, de Lange GJ (2003) Fluid-sediment interactions at Eastern Mediterranean mud volcanoes: a stable isotope study from ODP Leg 160. Earth Planet Sci Lett 212:377-391

11. Deyhle A, Kopf A (2002) Strong B enrichment and anomalous ^{11}B in pore fluids from the Japan Trench forearc. Mar Geol 183:1-15

12. Expedition 315 Scientists (2009a) Expedition 315 Site C0001. In: Kinoshita M, Tobin H, Ashi J, Kimura G, Lallemant S, Screaton EJ, Curewitz D, Masago H, Moe KT, Expedition 314/315/316 Scientists (ed) Proceedings of the Integrated Ocean Drilling Program, 314/315/316. Washington DC, pp 1–104

13. Expedition 315 Scientists (2009b) Expedition 315 Site C0002. In: Kinoshita M, Tobin H, Ashi J, Kimura G, Lallemant S, Screaton EJ, Curewitz D, Masago H, Moe KT, Expedition 314/315/316 Scientists (ed) Proceedings of the Integrated Ocean Drilling Program, 314/315/316. Washington DC, pp 1–75

14. Kinoshita M, Tobin H, Ashi J, Kimura G, Lallemant S, Screaton EJ, Curewitz D, Masago H, Moe KT (eds) (2009a) Expedition 316 Site C0004. In: Proceedings of the Integrated Ocean Drilling Program, 314/315/316 Integrated Ocean Drilling Program Management International, Inc, Washington DC. pp 1-107

15. Expedition 316 Scientists (2009b) Expedition 316 Site C0008. In: Kinoshita M, Tobin H, Ashi J, Kimura G, Lallemant S, Screaton EJ, Curewitz D, Masago H, Moe KT, Expedition 314/315/316 Scientists (ed) Proceedings of the Integrated Ocean Drilling Program, 314/315/316. Washington DC, pp 1–107

16. Gieskes JM, Gamo T, Brumsack H (1991) Chemical methods for interstitial water analysis aboard JOIDES Resolution. Ocean Drilling Program Texas A&M University Technical Note 15:1-60

17. Gieskes J, Mahn C, Day S, Martin JB, Greinert J, Rathburn T, McAdoo B (2005) A study of the chemistry of pore fluids and authigenic carbonates in methane seep environments: Kodiak Trench, Hydrate Ridge, Monterey Bay, and Eel River Basin. Chem Geol 220:329-345

18. Ginsburg G, Soloviev V, Matveeva T, Andreeva I (2000) Sediment grain-size control on gas hydrate presence, sites 994, 995, and 997. In: Paull CK, Matsumoto R, Wallace PJ, Dillon WP (ed) Proceedings of the Ocean Drilling Program, Scientific Results, 164: Ocean Drilling Program. College Station, pp 237–249

19. Greinert J, Bollwerk SM, Derkachev A, Bohrmann G, Suess E (2002) Massive barite deposits and carbonate mineralization in the Derugin Basin, Sea of Okhotsk: precipitation processes at cold seep sites. Earth Planet Sci Lett 203:165-180

20. Haese RR, Hensen C, de Lange GJ (2006) Pore water geochemistry of eastern Mediterranean mud volcanoes: Implications for fluid transport and fluid origin. Mar Geol 225:191-208

21. Harris RN, Schmidt-Schierhorn F, Spinelli G (2011) Heat flow along the NanTroSEIZE transect: Results from IODP Expeditions 315 and 316 offshore the Kii Peninsula, Japan. Geochem Geophys Geosyst 12:Q0AD16

22. Hensen C, Nuzzo M, Hornibrook E, Pinheiro LM, Bock B, Magalhães VH, Brückmann W (2007) Sources of mud volcano fluids in the Gulf of Cadiz - indications for hydrothermal imprint. Geochim Cosmochim Acta 71:1232-1248

23. Hesse R, Harrison WE (1981) Gas hydrates (clathrates) causing pore-water freshening and oxygen isotope fractionation in deep-water sedimentary sections of terrigenous continental margins. Earth Planet Sci Lett 55:453-462

24. Hiruta A, Snyder GT, Tomaru H, Matsumoto R (2009) Geochemical constraints for the formation and dissociation of gas hydrate in an area of high methane flux, eastern margin of the Japan Sea. Earth Planet Sci Lett 279:326-339

25. Ijiri A, Tsunogai U, Gamo T (2003) A simple method for oxygen-18 determination of milligram quantities of water using NaHCO3 reagent. Rapid Commun Mass Spectrom 17:1472–1478

26. Itai T, Kusakabe M (2004) Some practical aspects of an on-line chromium reduction method for D/H analysis of natural waters using a conventional IRMS. Geochem J 38:435-440

27. Kastner M, Elderfield H, Martin JB (1991) Fluids in convergent margins: what do we know about their composition, origin, role in diagenesis and importance for oceanic chemical fluxes? Philosophical Transactions of the Royal Society of London Series A - Mathematical Physical and Engineering Sciences 335:243-259

28. Kimura G, Moore GF, Strasser M, Screaton E, Curewitz D, Streiff C, Tobin H (2011) Spatial and temporal evolution of

the megasplay fault in the Nankai Trough. Geochem Geophys Geosyst 12:Q0A008

29. Kvenvolden KA, Kastner M (1990) Gas hydrates of the Peruvian outer continental margin. Proceedings of the Ocean Drilling Program, Scientific Results, 112: Ocean Drilling Program. College Station, pp 517–526

30. Lein AY, Pimenov NV, Savvichev AS, Pavlova GA, Vogt PR, Bogdanov YA, Sagalevich AM, Ivanov MV (2000) Methane as a source of organic matter and carbon dioxide of carbonates at a cold seep in the Norway Sea. Geochem Int 38:232-245

31. Maekawa T (2004) Experimental study on isotopic fractionation in water during gas hydrate formation. Geochem J 38:129-138

32. Maekawa T, Imai N (2000) Hydrogen and oxygen isotope fractionation in water during gas hydrate formation. In: Holder GD, Bishnoi PR (ed) Gas hydrates: challenges for the future, vol 912. Annals New York Academy Sciences, pp 452–459

33. Magaritz M, Gat JR (1981) Review of the natural abundance of hydrogen and oxygen isotopes. In: Gat JR, Gonfiantini R (eds) Stable isotope hydrology - deuterium and oxygen-18 in the water cycle, International Atomic Energy Agency, Vienna. pp 85-102

34. Manheim FT (1968) Disposable syringe techniques for obtaining small quantities of pore water from unconsolidated sediments. J Sediment Petrol 38:666-668

35. Martin JB, Kastner M, Henry P, Le Pichon X, Lallement S (1996) Chemical and isotopic evidence for sources of fluids in a mud volcano field seaward of the Barbados accretionary wedge. J Geophys Res 101:20325-20345

36. Mazurenko LL, Soloviev VA, Gardner JM, Ivanov MK (2003) Gas hydrates in the Ginsburg and Yuma mud volcano sediments (Moroccan Margin): results of chemical and isotopic studies of pore water. Mar Geol 195:201-210

37. Miyajima T, Yamada Y, Handa YT, Yoshii K, Koitabashi T, Wada E (1995) Determining the stable-isotope ratio of total dissolved inorganic carbon in lake water by GC/C/IRMS. Limnol Oceanogr 40:994-1000

38. Murray RW, Miller DJ, Kryc KA (2000) Analysis of major and trace elements in rocks, sediments, and interstitial waters by

inductively coupled plasma-atomic emission spectrometry (ICP-AES). Ocean Drilling Program Texas A&M University Technical Note 29:1-27

39. Naehr TH, Stakes DS, Moore WS (2000) Mass wasting, ephemeral fluid flow, and barite deposition on the California continental margin. Geology 28:315-318

40. Park J-O, Kodaira S (2012) Seismic reflection and bathymetric evidences for the Nankai earthquake rupture across a stable segment-boundary. Earth Planets Space 64:299-303

41. Park J-O, Tsuru T, Kodaira S, Cummins PR, Kaneda Y (2002) Splay fault branching along the Nankai subduction zone. Science 297:1157-1160

42. Parsons B, Sclater JG (1977) An analysis of the variation of ocean floor bathymetry and heat flow with age. J Geophysical Rearch 82:6B0585

43. Paull CK, Hecker B, Commeau R, Freeman-Lynde RP, Neumann C, Corso WP, Golubic S, Hook JE, Sikes E, Curray J (1984) Biological communities at the Florida escarpment resemble hydrothermal vent taxa. Science 226:965-967

44. Reid JL (2009) On the world-wide circulation of the deeper waters of the world ocean: Scripps Institution of Oceanography, UC San Diego

45. Reitz A, Haeckel M, Wallmann K, Hensen C, Heeschen K (2007) Origin of salt-enriched pore fluids in the northern Gulf of Mexico. Earth Planet Sci Lett 259:266-282

46. Savin SM, Epstein S (1970) The oxygen and hydrogen isotope geochemistry of ocean sediments and shales. Geochim Cosmochim Acta 34:48-63

47. Screaton EJ, Kimura G, Curewitz D (2009) Expedition 316 summary. In: Kinoshita M, Tobin H, Ashi J, Kimura G, Lallemant S, Screaton EJ, Curewitz D, Masago H, Moe KT (eds) Proceedings of the Integrated Ocean Drilling Program Expeditions, 314/315/316, Integrated Ocean Drilling Program Management International, Inc, Washington DC.

48. Sheppard SMF, Gilg HA (1996) Stable isotope geochemistry of clay minerals. Clay Miner 31:1-24

49. Sloan EDJ, Koh CA (2008) Clathrate hydrates of natural gases

(third edition). CRC, Taylor & Francis Group, Boca Raton.

50. Suess E, Ritger SD, Moore JC, Jones ML, Kulm LD, Cochrane GR (1985) Biological communities at vent sites along the subduction zone off Oregon. Bulletin of the Biological Society of Washington 6:474-484

51. Teichert BMA, Torres ME, Bohrmann G, Eisenhauer A (2005) Fluid sources, fluid pathways and diagenetic reactions across an accretionary prism revealed by Sr and B geochemistry. Earth Planet Sci Lett 239:106-121

52. Toki T, Tsunogai U, Gamo T, Kuramoto S, Ashi J (2004) Detection of low-chloride fluids beneath a cold seep field on the Nankai accretionary wedge off Kumano, south of Japan. Earth Planet Sci Lett 228:37-47

53. Toki T, Maegawa K, Tsunogai U, Kawagucci S, Takahata N, Sano Y, Ashi J, Kinoshita M, Gamo T (2011) Gas chemistry of pore fluids from Oomine Ridge on the Nankai accretionary prism. In: Ogawa Y, Anma R, Dilek Y (eds) Accretionary prisms and convergent margin tectonics in the Northwest Pacific Basin, vol. 8, Springer, Heidelberg. pp 247-262

54. Toki T, Uehara Y, Kinjo K, Ijiri A, Tsunogai U, Tomaru H, Ashi J (2012) Methane production and accumulation in the Nankai accretionary prism: Results from IODP Expeditions 315 and 316. Geochem J 46:89-106

55. Toki T, Higa R, Tanahara A, Ijiri A, Tsunogai U, Ashi J (2013) Origin of pore water in Kumano mud volcanoes (in Japanese with English abstract). Chikyukagaku (Geochemistry) 47:221-236

56. Tomaru H, Matsumoto R, Lu H, Uchida T (2004) Geochemical process of gas hydrate formation in the Nankai Trough based on chloride and isotopic anomalies in interstitial water. Resour Geol 54:45-51

57. Tsunogai U, Wakita H (1995) Precursory chemical changes in ground water: Kobe earthquake, Japan. Science 269:61-63

58. Tsunogai U, Yoshida N, Gamo T (2002) Carbon isotopic evidence of methane oxidation through sulfate reduction in sediment beneath cold seep vents on the seafloor at Nankai Trough. Mar Geol 187:145-160

59. Ussler W, Paull CK (1995) Effects of ion exclusion and isotopic

fractionation on pore water geochemistry during gas hydrate formation and decomposition. Geo-Mar Lett 15:37-44

60. Vrolijk P, Chambers SR, Gieskes JM, O'Neil JR (1990) Stable isotope ratios of interstitial fluids from the northern Barbados accretionary prism, ODP Leg 110. In: Moore JC, Mascle A, Taylor E, Underwood MB (ed) Proceedings of the Ocean Drilling Program, Scientific Results, 110: Ocean Drilling Program. College Station, pp 189–205

61. Vrolijk P, Fisher A, Gieskes J (1991) Geochemical and geothermal evidence for fluid migration in the Barbados accretionary prism (ODP leg 110). Geophys Res Lett 18:947-950

62. Wallmann K, Linke P, Suess E, Bohrmann G, Sahling H, Schlüter M, Dählmann A, Lammers S, Greinert J, Von Mirbach N (1997) Quantifying fluid flow, solute mixing, and biogeochemical turnover at cold vents of the eastern Aleutian subduction zone. Geochim Cosmochim Acta 61:5209-5219

63. Yeh HW (1980) D/H ratios and late-stage dehydration of shales during burial. Geochim Cosmochim Acta 44:341-352

64. You CF, Gieskes JM (2001) Hydrothermal alteration of hemi-pelagic sediments: experimental evaluation of geochemical processes in shallow subduction zones. Appl Geochem 16:1055-1066

65. You C-F, Gieskes JM, Chen RF, Spivack AJ, Gamo T (1993a) Iodide, bromide, manganese, boron, and dissolved organic carbon in interstitial waters of organic carbon-rich marine sediments: observations in the Nankai accretionary prism. In: Hill IA, Taira A, Firth JV (ed) Proceedings of the Ocean Drilling Program, Scientific Results, 131: Ocean Drilling Program. College Station, pp 165–174

66. You CF, Spivack AJ, Smith JH, Gieskes JM (1993) Mobilization of boron in convergent margins: Implications for the boron geochemical cycle. Geology 21:207-210

67. You CF, Castillo PR, Gieskes JM, Chan LH, Spivack AJ (1996) Trace element behavior in hydrothermal experiments: Implications for fluid processes at shallow depths in subduction zones. Earth Planet Sci Lett 140:41-52

Metabolic Flux and Transcriptional Analysis Elucidate Higher Butanol/ Acetone Ratio Feature in Abe Extractive Fermentation by Clostridium acetobutylicum Using Cassava Substrate

Xin Li, Zhi-Gang Li, and Zhong-Ping Shi

Key Laboratory of Industrial Biotechnology, Ministry of Education, Jiangnan University, Wuxi 214122, People's Republic of China

ABSTRACT

Background

In acetone-butanol-ethanol (ABE) fermentation by *Clostridium acetobutylicum* ATCC 824 using corn-based substrate, the solvents are generally produced at a ratio of 3:6:1 (A:B:E, w/w).

Results

A higher butanol/acetone ratio of 2.9:1 was found when cassava was used as the substrate of an in-situ extractive fermentation by *C. acetobutylicum*. This ratio had a 64% increment compared to that on corn-based substrate. The metabolic flux and (key enzymes) genes transcriptional analysis indicated that weakened metabolic fluxes in organic acids (especially butyrate) formation and re-assimilation pathways, which associated with lower *buk* and *ctfAB* transcriptional levels, contributed to higher butanol and lower acetone production rate in fermentations on cassava. Moreover, NADH generation was enhanced under the enriched reductive environment of using cassava-based substrate, which converted more carbon flux to butanol synthesis pathway, also leading to a higher ratio of butanol/acetone. To further increase butanol/acetone ratio, tiny amount of electron carrier, neutral red was supplemented into cassava-based substrate at 60 h when butonal production rate reached maximal level. However, neutral red addition enhanced NADH production, followed with strengthening the metabolic fluxes of organic acids formation/re-assimilation pathways, resulted in unchanged in butanol/acetone ratio.

Conclusions

A further increase in butanol/acetone ratio could be realized when NADH regeneration was enhanced and the metabolic fluxes in organic acids formation/reutilization routes were controlled at suitably low levels simultaneously.

BACKGROUND

Clostridium acetobutylicum, a Gram-positive, spore-forming, and obligate anaerobe, has the ability to produce solvents with renewable biomasses including acetone, butanol, and ethanol [1]. In acetone-butanol-ethanol (ABE) fermentation by *C. acetobutylicum* using corn-based substrate, the solvents are generally produced at a ratio of 3:6:1 (A/B/E, *w/w*). Among these solvents, butanol has the most attraction since it has been considered as a high-performance biofuel, as well as an important platform chemical. However, high substrate (such as corn) price and too much purification cost due to very low solvent concentrations are the two major factors impacting on economics of butanol production [2]. A cost sheet from an ABE fermentation plant using corn indicated that the substrate price accounted for up to 79% of the overall production cost, while energy cost consumed in product distillation almost contributed the rest 14% of the entire cost [3]. Therefore, seeking cheaper feedstock and increasing butanol ratio in total solvents have become the major challenges for the economic viability of ABE fermentation.

Some low-priced agriculture residues for example corn fiber and wheat bran have been used in ABE fermentation by clostridia, but butanol concentration and productivity in these fermentations are much lower than those in corn-based fermentation [4]-[7]. Cassava, a non-grain and high starch content crop, is recognized as an economical and practical substrate for industrial fermentation. In some previous studies, cassava was successfully used instead of corn as substrate in ABE fermentation by *C. acetobutylicum*, and higher butanol/acetone ratios were observed when fermenting on cassava as compared with the same procedures using corn [8],[9]. In these cases, the online monitored parameters (pH, ORP, H_2/CO_2 ratio, etc.) and organic acid formation/reassimilation patterns were quite different from those on corn-based substrate. But the mechanism of higher butanol ratios in cassava-based fermentation has still not been illustrated clearly, for the activities of key enzymes *in vivo* were difficult or impossible to be measured, caused by hardly separating cells from mixed solid residues of substrate.

Some efforts have investigated the special features presented in ABE fermentation by *C. acetobutylicum*, using either metabolic flux analysis or transcriptional analysis [10]-[12]. Metabolic flux analysis is a systematic approach developed to evaluate each individual reaction rate within a metabolic network. Investigation on genetic transcriptional levels directly correlates the activities of relevant enzymes. To solve the problems mentioned above, the methods of metabolic flux analysis was combined with genes (key enzymes) transcriptional measurements to explore the mechanism of higher butanol/acetone ratio feature in cassava-based fermentation. Traditional batch process is still the most commonly used operation mode in industrial ABE fermentations. However, it is suffered with severe butanol end-inhibition leading to a short fermentation period, so that interpreting many attractive phenomena becomes difficult. By contrast, in-situ extractive fermentation could relieve butanol inhibitory effect to improve fermentation productivity and to prolong fermentation time [13]-[15]. The in-situ extractive fermentation technique is not widely used in industrial ABE fermentation because of the high extractant cost and operation complexity. However, it could be used as an important prototype for investigating various characteristics of ABE fermentation, and guiding the optimal operation ways of ABE traditional fermentation. Among various in-situ fermentation extractants for butanol, oleyl alcohol has been recognized as the best one because of its non-toxicity to cell growth and high butanol extraction coefficient [16]. In this study, ABE extractive fermentations by *C. acetobutylicum* ATCC 824 were conducted in a 7-L anaerobic fermentator, under the conditions of using different biomass substrate (corn or cassava). Combinational analysis of metabolic flux distribution and gene transcriptional levels were carried out to find out the variations in intracellular carbon distributions and transcriptional levels when using corn- or cassava-based substrate. All these efforts aimed to clarify the mechanism of higher butanol/acetone ratio obtained when using cassava-based medium and explore the optimal operation way for traditional ABE fermentation characterized with high butanol/acetone ratio.

METHODS

Microorganism

C. acetobutylicum ATCC 824 was used in this study. The strain was maintained as spore suspension in 5% corn meal medium at 4°C. The methods of inoculation and pre-culture followed those described in the literatures [17],[18].

Substrate (Media) Preparation and In-situ Fermentative Extractant Pretreatment

The corn flour (raw starch content about 50% w/w) was obtained at local market and cassava powder (raw starch content about 65% to 70% w/w) was provided by Henan Tianguan Fuel Ethanol Co. Ltd., Nanyang, China. The media were pretreated by adding certain amount of α-amylase (8 U/g-corn or cassava, heated in boiling water bath for 45 min) and then glucoamylase (120 U/g-corn or cassava, heated at 62°C for 60 min). Subsequently, the viscosity-reduced media were autoclaved at 121°C for 20 min. Oleyl alcohol (Tokyo Kasei Co. Ltd., Tokyo, Japan) was used as the extractant for in-situ extractive fermentation. Oleyl alcohol was either sterilized at 121°C for 20 min or directly used without sterilization, and then added into the fermentor. When using cassava as substrate, the concentrated yeast extract solution was sterilized at 115°C for 30 min, and then pumped into the broth upon requirement since it has been revealed in the previous study that yeast extract could promote the phase shift in ABE fermentation with cassava substrate [9]. Neutral red was dissolved in sterilized water and pumped into the broth at 60 h.

Fermentation Method and Condition

Seed culture was carried out in 100-mL anaerobic fermentation bottles using corn as the substrate. The initial corn or cassava meal content for extractive fermentations was 30% or 25% (w/v). The fermentations were conducted in a 7-L static fermentor (Baoxing Bioengineering Co., Shanghai, China) equipped with pH and ORP (oxidative-reductive potential) electrodes and a manual adjusted pressure unit. A temperature-controllable water bath (MP-10, Shanghai Permanent Science and Technology, Co., Shanghai, China) was used to circulate hot water into the coil pipes settled inside the fermentor to maintain broth temperature at 37°C. The fermentation medium loading volume ranged from 1.8 to 2.5 L, and equivalent volume of oleyl alcohol was added to ensure a 1:1 oil/broth volumetric ratio. N_2 was sparged into the extractant reservoir for 10 min to remove residual oxygen. 10% (v/v) inoculum was transferred into the fermentor and then N_2 was also sparged into broth for 10 min. The oxygen-free oleyl alcohol was poured into the fermentor using a peristaltic pump after inoculation. The initial pressure inside the fermentor was controlled at about 0.02 MPa (N_2) to strictly maintain the anaerobic condition. The pressure gradually increased since fermentation started and self-generated gas began to evolve. The pressure was then controlled in a range of 0.030 to 0.055 MPa throughout fermentation. Agitation was occasionally adopted for a short time (5 min, 400 rpm) to promote butanol diffusion from aqueous phase into extractant phase.

Analytical Methods

The measurements of concentrations of solvents, organic acids and reducing sugar (glucose) were the same as those described in our previous reports [17],[18]. On account of the volume ratio of aqueous to organic phase being just 1:1 in extractive fermentation, the total concentration of butanol (or acetone) was the sum of butanol (or acetone) in broth and in extractant. H_2/CO_2 ratio in exhaust gas was determined using the same method reported in our previous work [9]. The concentration of important intermediate, butyraldehyde in broth was determined by a gas chromatography (Shimadzu GC-2010, Kyoto, Japan) with flame ionization detector and DB-23 capillary column (60 m × 0.25 mm ID × 0.32 μm, Agilent, Sta. Clara, CA, USA). The condition was described as follows: nitrogen was used as the carrier gas at a velocity of 1.2 mL/min; the hydrogen and air flow rates were 47 and 400 mL/min, respectively; injector temperature was operated at 200°C, and detector temperature at 250°C; the initial temperature stayed at 40°C for 5 min, and then raised at the velocity of 10°C per min until arriving to 180°C, and finally stayed for another 5 min.

Metabolic Flux Analysis

Metabolic flux analysis involves the calculation (or estimation) of *in vivo* fluxes from substrate and product data, by using a system of linear equations developed from reaction stoichiometry [19]. For the purpose of metabolic flux distribution analysis, a simplified metabolic reaction model (MR) was developed for butanol synthesis by *C. acetobutylicum* ATCC 824. This simplified model (Figure 1) covered the basic reactions and occurred in the glycolysis pathway, organic acids formation/reutilization routes, solvent synthesis branches, and the electron transport shuttle system. Cell growth and ATP synthesis were not included in the model, since the metabolic flux distribution analysis was conducted in the solventogenic phase where cell growth almost ceased and ATP demand was less [20].

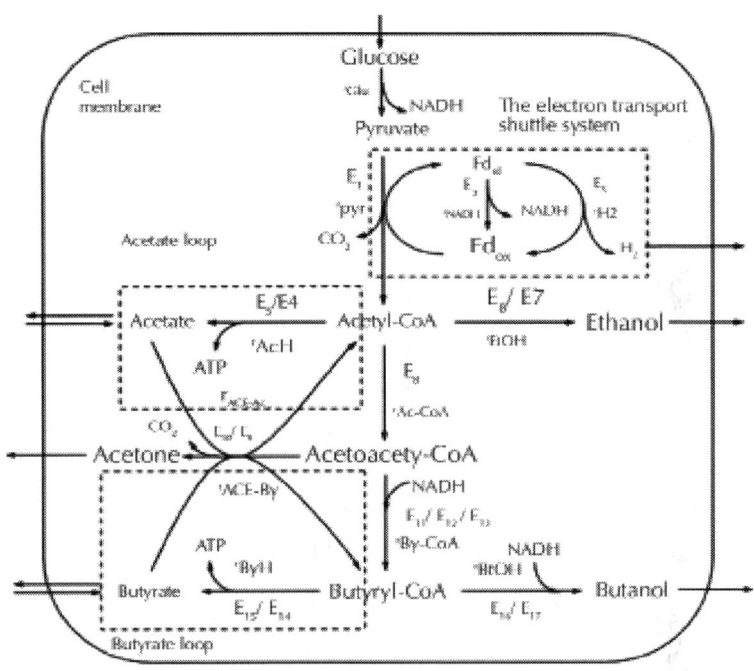

Figure 1: The basic metabolic pathway of acetone and butanol biosynthesis by*Clostridium acetobutylicum*ATCC824.

As shown in Figure 1 and the Appendix, the MR model contained 19 metabolic reaction rates ($k = 19$). Among the rates, seven extracellular substance rates were measurable ($m = 7$) including rates of glucose consumption, organic acid formation or re-assimilation, solvents synthesis, and hydrogen evolution. As shown in the Appendix, there were a total of 13 substances ($n = 13$) covering substrates, products and intermediate metabolites, and thus 13 mass balance equations were available. Thus, this MR model is an overdetermined system

($n = 13 > k\text{-}m = 12$). All of the unknown reaction rates could be optimally determined using the measurable rate data and the stoichiometric coefficients of the metabolic reaction matrix with the aid of the calculation package embedded in Matlab (Ver. R2010b, MathWorks Inc., Natick, MA, USA) [21]. The following treatments were applied in the network model calculation: (1) glucose was calculated in the model as single carbon resource since it was the most preferred for strain and highest percentage in these complex mediums; (2) glucose consumption rates were normalized as 100 mmol/(L·h), and the other measurable rates were recalculated using the above (glucose) normalization coefficient; (3) pseudo-steady state assumption was adopted for intracellular intermediate metabolites.

RNA Purification, cDNA Synthesis, and Real-time Fluorescence Quantitative PCR Analysis

Total bacterial RNA was extracted using Trizol Plus RNA Purification Kit (Invitrogen™, Life Technologies Corp., Grand Island, NY, USA). Before starting RNA extraction, all the samples were percolated through filter paper to remove cassava/corn residues. 1 mL Trizol® Reagent was added to wet cell for lysing cell and disassociating protein from nucleic acid. Then, 0.2 mL of chloroform was supplied into the homogenate liquid following with a vigorous shaking, in order to extract the protein out of aqueous phase. After transferring the aqueous phase to a new tube, 100% isopropanol (0.5 mL) was added to separate RNA out of the aqueous phase. It must be noted that using the appropriate precautions to avoid RNase contamination when preparing and handling RNA. Total RNA was used as the template to synthesize cDNA, and then cDNA products were amplified by the method of real-time fluorescence quantitative PCR with the primers listed in Table 1. The following PCR conditions were adopted: an initial denaturation step at 95°C for 10 min, followed by an amplification and quantification program repeating for 40 cycles (95°C for 10 s, 60°C for 60 s with a single fluorescence measurement), and a melting curve program (a continuous fluorescence measurement raising temperature from 60°C to 95°C with a slow heating unit).

Table 1: Primer sequences used in the real-time fluorescence quantitative PCR

Gene	Prime sequences	Fragment size (bp)
16S RNA	F:5'CTGGACTGTAACTGACGCTGA3'	80
	R:5'CGTTTACGGCGTGGACTAC3'	
ctfAB	F:5'CAGAAAACGGAATAGTTGGAATG3'	151
	R:5'TGACCACCACGGATTAGTGAA3'	
adhE	F:5'GTTTTGGCTATGTATGAGGCTGA3'	241
	R:5'CAAGCGTGAAAGAAGGTGGTAT3'	
bdhB	F:5'ACGCTTCTGCCATTCTATCC3'	175
	R:5'ATTGCGGCACATCCAGATA3'	
ask	F:5'GTATGGGATTTACTCCTCTTGG3'	63
	R:5'CTGGGTCCATATCTCCACTTC3'	
buk	F:5'TCCGCCTTTGCCGTTTA3'	194
	R:5'ACATGGGTGGAGGTACTTCAGT3'	

Li *et al.*

Li *et al. Bioresources and Bioprocessing* 2014 **1**:13, doi:10.1186/s40643-014-0013-9

RESULTS

Fermentation Performances Comparison when Using Corn or Cassava as Substrate

Figure 2 shows the extractive fermentation performances (with oleyl alcohol as extractant) when using either corn or cassava as substrate. The change patterns of pH and total gas production yields no significant distinction (Figure 2a). However, ORP and H_2/CO_2 ratio for the two cases (Figure 2b) are quite different. ORP in cassava-based broth reached −520 mV in the solventogenic phase of fermentation, which was visibly lower than that of corn-based broth (−470 mV). H_2/CO_2 ratio stayed at a low level of 0.2 to 0.7 when culturing with cassava-based substrate, while fluctuating in the range of 0.6 to 2.5 in corn-based fermentation.

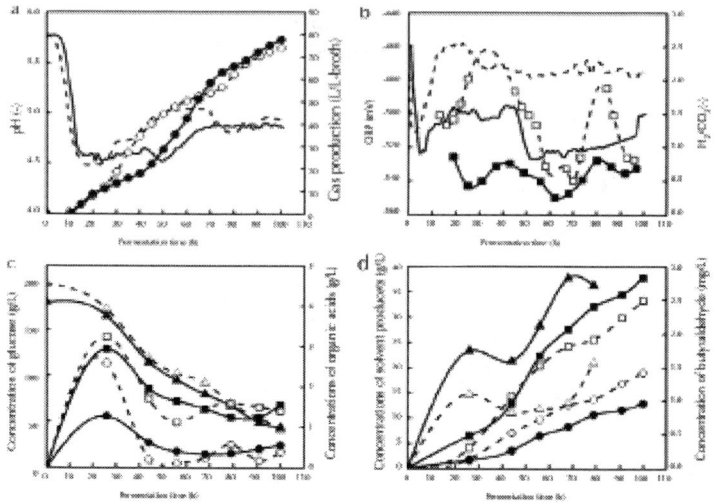

Figure 2: Performance comparisons of butanol fermentations using corn-(open symbols and broken lines) and cassava-based substrates (filled symbols and solid lines). (a) pH (line) and gas production (circle). (b) ORP (line) and H_2/CO_2 ratio (square). (c)Acetate (square), butyrate (circle), and glucose concentrations (triangle).(d) Total butanol concentration (square), total acetone concentration (circle), and butyraldehyde concentration in broth (triangle).

The maximum butyrate accumulation in cassava broth was 1.2 g/L, only half of which in corn broth. Furthermore, after entering solvent genic phase, acetate and butyrate re-assimilation rates were significantly slower in cassava-based fermentation. The final solvent concentrations (including butanol and acetone in both extractive and aqueous phase. ethanol was not accounted for due to low accumulation) reached a total level of 50 g/L in both cases, in which butanol/acetone ratio was 2.87 using cassava-based substrate and 1.75 with corn. All these distinctions, particularly butanol/acetone and H_2/CO_2 ratios, were important in understanding the optimal regulation for ABE fermentation. Their mechanisms should be properly interpreted.

Metabolic Flux Analysis in ABE Extractive Fermentations with Corn and Cassava

High butanol/acetone ratio is desirable for ABE fermentation. It reached nearly 2.9 when using cassava-based substrate, a 64% enhancement compared to corn. To understand this result, metabolic flux analysis was conducted to explore the mechanisms.

As shown in Figure 1, the metabolic fluxes of butanol production by C. actobutylicum ATCC 824 mainly includes the metabolic reaction rates of r_{Ac-CoA} and r_{By-CoA} in central butanol synthesis route after acetyl-CoA node, r_{AcH} and r_{ByH} in acetate/butyrate formation routes, r_{ACE-Ac} and r_{ACE-By} in acetone synthesis branch, r_{BtOH} in butanol production branch, r_{EtOH} in ethanol formation branch and r_{NADH} (NADH generation rate) in the electron transport shuttle system. The metabolic fluxes during the solventogenic phase (40 to 100 h) are depicted in Figure 3. As shown in the figure, when fermenting on corn-based substrate, acetate and ethanol formation fluxes (r_{AcH} and r_{EtOH}) are much higher than those on cassava-based substrate throughout the solventogenic phase, resulting in a less carbon flux in the central butanol routes (r_{Ac-CoA}). Moreover, butyrate formation flux of r_{ByH} when fermenting corn-based substrate is also significantly higher. The butyrate formation (r_{ByH}) flux is inversely associated with the flux in butanol production branch (r_{BtOH}) because they compete for carbon resource at butyryl-CoA node. Therefore, a higher r_{ByH} (in the case of fermenting corn-based substrate) led to a lower flux in butanol synthesis branch (r_{BtOH}). Since acetone formation branch flux is the sum of r_{ACE-Ac} and r_{ACE-By} (shown in the Appendix), when fermenting cassava-based substrate, both r_{ACE-Ac} and r_{ACE-By} declined sharply after 55 h, resulting in a lower acetone synthesis flux. In addition, the reductive power NADH, an extremely important cofactor in butanol synthesis, also may have contribution to the extent of butanol/acetone ratio. NADH is mainly produced from two pathways, the glycolysis and the electron transport shuttle system. In this study, the glucose consumption amount and rate using two different substrates had nearly the same change patterns (shown in Figure 2c), which meant that the NADH produced from the glycolysis were equivalent between fermentations on corn and cassava. However, the NADH generated from the electron transport shuttle system had visibly different production rates in two cases. As shown in Figure 3, r_{NADH} stays at high levels during most period of the solventogenic phase when using cassava-based substrate. Correspondently, fluxes in the NADH-dependent pathways of butyryl-CoA and butanol formations (r_{By-CoA} and r_{BtOH}) are largely enhanced in the case of fermenting cassava. In order to verify the NADH balance, the difference values between NADH production ($2 \times r_{gly} + r_{NADH}$) and utilization ($2 \times r_{EtOH} + 2 \times r_{By-}$

$_{CoA} + 2 \times r_{BtOH}$) were calculated. The result indicated that the values were almost in the range from 5 to −5, accounting for less than 5% of NADH production (or utilization), which confirmed the reliability of metabolic flux analysis. All of the metabolic flux analysis result supports the feature of high butanol/acetone ratio when adopting cassava substrate.

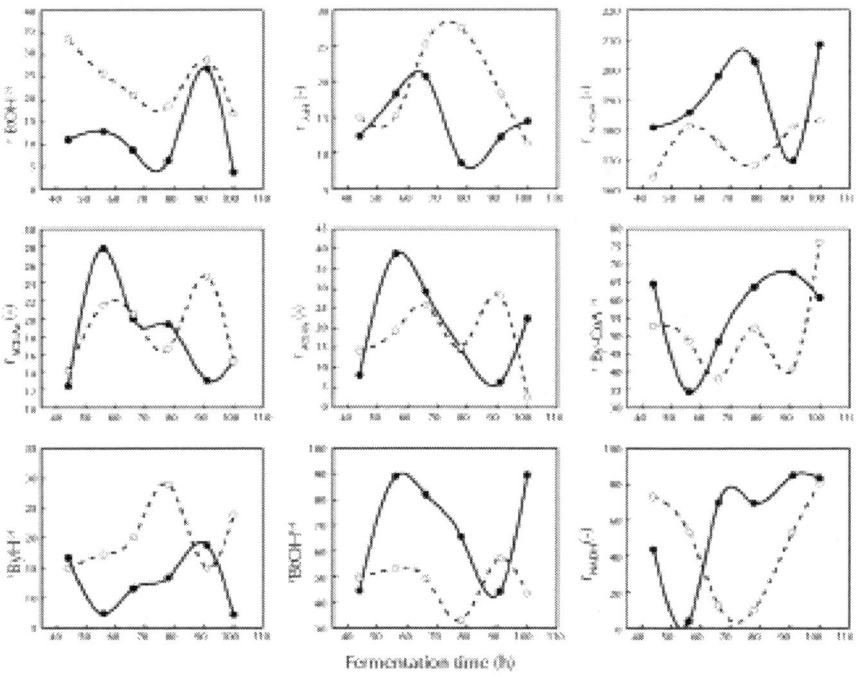

Figure 3: Metabolic flux analysis of *Clostridium acetobutylicum* ATCC824 when using corn- and cassava-based substrates. Corn-based substrates (open symbols and broken lines); cassava-based substrates (filled symbols and solid lines).

Gene Transcriptional Analysis for ABE Extractive Fermentations with Corn and Cassava

The real-time fluorescence quantitative PCR was also conducted to further interpret the mechanism of high butanol/acetone ratio in cassava-based fermentation. According to butanol synthesis map shown in Figure 1, acetate kinase and butyrate kinase responsible for acetate and butyrate synthesis respectively are encoded by the genes of *ask* and *buk*; CoA-transferase encoded by *ctfAB* is in charge of acid reutilization coupling with acetone formation; butyraldehyde dehydrogenase and butanol dehydrogenase encoded by *adhE* and *bdhB* are the key enzymes in the final two steps of butanol production. The results of gene transcriptional analysis are depicted in Figure 4. Variation in the transcriptional level of *ask* during the solventogenic phase was limited regardless of the substrate types. However, the transcriptional level of *buk* decreased significantly in fermentation using cassava-based substrate, while staying rather high in corn-based fermentation. Transcriptional level of *ctfAB* reached the maximal level at 79 h when using corn-based substrate, which was approximately sevenfold higher than that in cassava-based fermentation. Likewise, the maximal values of *adhE* and *bdhB* appeared at 79 h when fermenting on corn, which were respectively eightfold and twofold higher than those on cassava.

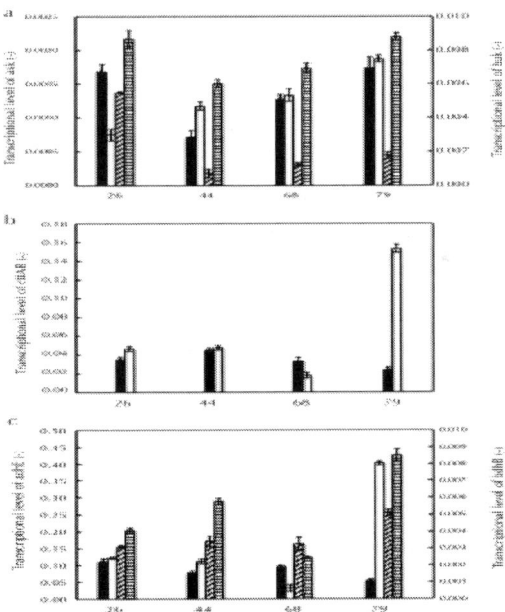

Figure 4: Changes in transcriptional levels of key genes in corn- and cassava-based fermentations. Corn-based fermentation (white and parallel shadow

bars); cassava-based fermentation (black and slashed shadow bars). (a) Transcriptional levels of *ask* (black and white) and *buk* (shadow). (b) Transcriptional level of *ctfAB*. (c) Transcriptional levels of *adhE* (black and white) and *bdhB* (shadow).

Attempt of Further Butanol/Acetone Ratio Enhancement by Adding neutral Red in ABE Fermentation with Cassava

Neutral red, an electron carrier, could force electron flow direction change. Adding a tiny amount of neutral red could weaken hydrogen formation which originated from the electron shuttle transport system, leading to an enhancement in NADH production. In order to further increase the butanol/acetone ratio, neutral red (0.1% w/v-broth) was added into cassava broth at 60 h, when butanol production was in a relatively high rate. As expected, r_{NADH} was visibly elevated and then sustained at higher levels after neutral red addition (Figure 5d). Moreover, an increment of butyraldehyde concentration was observed, which suggests that butanol synthesis pathway was enhanced. In fact, butanol production was truly increased and reached 43 g/L (Figure 5c). However, butanol/acetone ratio was not improved by adding neutral red. Metabolic flux analysis indicated that the rates of organic acid formation and re-assimilation were promoted after neutral red addition, although the concentrations of acids in broth had not an apparent change. Acetone production was consequently promoted. Therefore, simultaneous increase in both butanol and acetone production led to an unchanged butanol/acetone ratio in this case. Table 2 summarizes the fermentation performance under different runs.

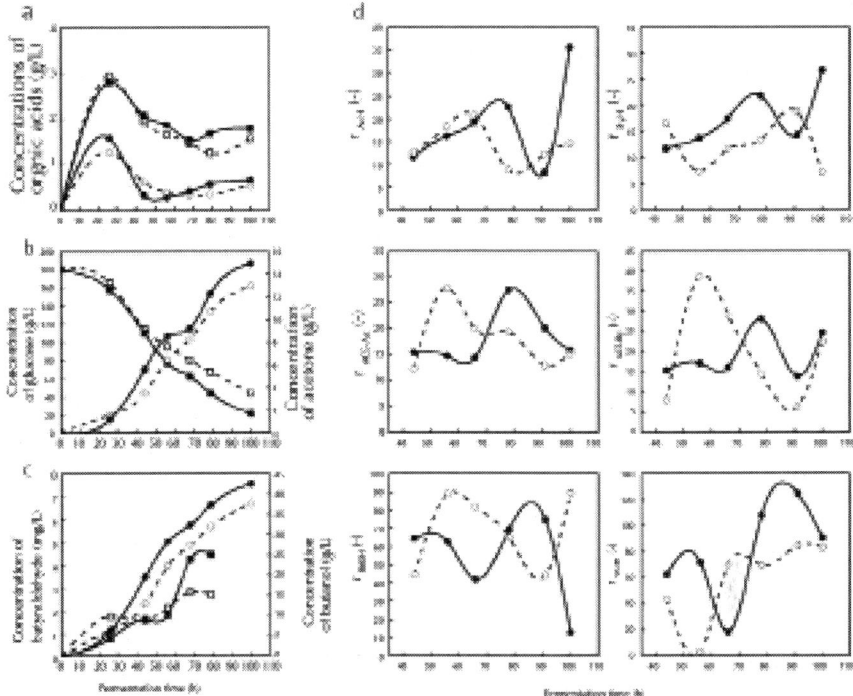

Figure 5: Fermentation performances and metabolic flux analysis of*Clos-tridium acetobutylicum*ATCC824 when using cassava-based substrate. The substrate is either with (filled symbols and solid lines) or without neutral red addition (open symbols and broken lines). (a) Acetate (square) and butyrate concentrations (circle). (b)Glucose (square) and total acetone concentrations (circle). (c) Butyraldehyde concentration in broth (square) and total butanol concentration (circle). (d) Metabolic fluxes in cell supplemented with or without neutral red.

Table 2: Fermentation performance under various operation modes with corn and cassava substrates

Fermentation characteristics	Substrate		
	Corn	Cassava	Cassava + Neutral red
Number of experiments	2	2	2

Fermentation time (h)	100	100	100
Total butanol (g/L)	33.26±2.61	37.48±3.11	42.60±2.02
Total acetone (g/L)	18.98±1.39	13.06±0.93	14.59±0.60
Butanol/ acetone ratio (−)	1.75±0.01	2.87±0.03	2.92±0.02
Acetate (g/L)			
Max.	3.20±0.60	2.94±0.27	2.80±0.03
Final	1.40±0.22	1.24±0.34	1.60±0.08
Butyrate (g/L)			
Max.	2.58±0.34	1.27±0.16	1.44±0.28
Final	0.39±0.06	0.41±0.12	0.51±0.15
Gas production (L/L-broth)	80.00±6.13	72.37±5.37	64.86±0.11
Butanol productivity (g/L/h)	0.33±0.02	0.37±0.03	0.42±0.02

Li et al.

Li et al. Bioresources and Bioprocessing 2014 **1**:13, doi:10.1186/s40643-014-0013-9

Figure 6 compares the transcriptional levels of genes encoding the key enzymes in butanol synthesis with/without neutral red supplement. After neutral red addition, the transcriptional levels of *ctfAB* and *adhE*, responsible for acetone and butanol synthesis respectively, were apparently enhanced. This result was basically in accordance with the metabolic flux analysis results mentioned above.

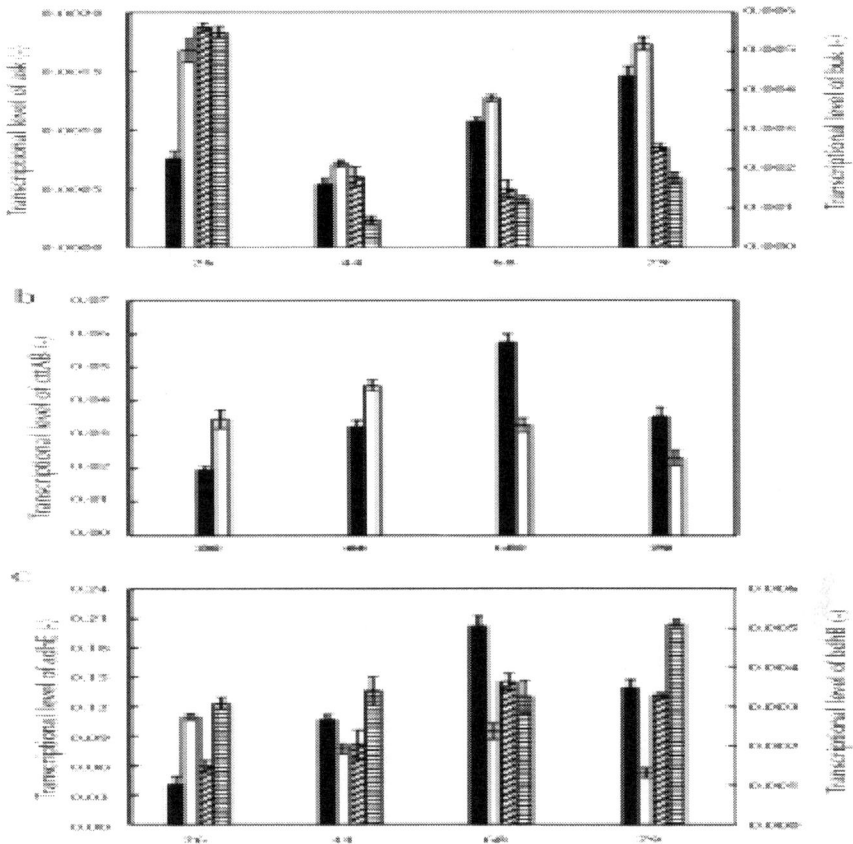

Figure 6: Changes in transcriptional levels of key genes in fermentations using cassava-based substrate. The substrate is either with (black and slashed shadow bars) or without neutral red addition (white and parallel shadow bars). (a) Transcriptional levels of *ask* (black and white) and *buk* (shadow). (b) Transcriptional level of *ctfAB*. (c) Transcriptional levels of *adhE* (black and white) and *bdhB* (shadow).

DISCUSSION

The butanol/acetone ratio reached a much higher level of 2.87 in extractive fermentation with cassava substrate, a 64% increment as compared to using corn substrate (Table 2). It was close to the reported ratio obtained in fermentations by a hyper-butanol strain of *C. acetobutylicum* EA 2018 with corn-based medium [22]. The preliminary analysis to fermentation performances showed that ORP and H_2/CO_2 ratio were both at low levels when using cassava (Figure 2). ORP could be considered as a comprehensive index reflecting pH, dissolved oxygen, reductive potential of compounds dissolved in medium [23]. The ABE fermentation required an anaerobic environment so that dissolved oxygen in broth could be ignored. Change patterns of pH were basically similar when using different substrates. Thus, the lower ORP suggested that cassava-based medium was rich in reductive substances. On the other hand, H_2 and CO_2 are the two major components in the exhaust gas emitted by clostridia. H_2 is generated from the electron transport shuttle system via the reaction of $2H^+ + 2e^- \rightarrow H_2$ catalyzed by hydrogenase [24]. CO_2 is mainly produced in reaction of Pyruvate $\rightarrow$ Acetyl-CoA associated with formation of reductive ferredoxin, the electron donor for hydrogen or NADH generation [25]. Therefore, lower H_2/CO_2 ratio implied that more electron flows were distributed to NADH production in the electron transport shuttle system.

Based on the preliminary analysis results, metabolic flux and gene transcriptional analysis were conducted to verify the assumption and to elucidate the mechanism about high butanol/acetone ratio in cassava-based fermentation. The metabolic flux analysis revealed that NADH was truly generated more under the cassava-based environment (Figure 3), which was consistent with the preliminary analysis results. It should be addressed that the genes (*adhE* and *bdhB*) regulating butanol synthesis had higher transcriptional levels under corn-based environment, but more butanol production was obtained in cassava-based fermentation. This fact demonstrated that NADH generation rate is one of dominated factors controlling butanol/acetone ratio in ABE fermentations by *C. acetobutylicum*. Besides, metabolic flux of butyrate closed-loop (r_{ByH} and r_{ACE-By}) in cassava-based fermentation was largely weakened after shifting into solventogenic phase. Correspondingly, transcriptional levels of *buk* and *ctfAB* (responsible for butyrate formation and re-assimilation respectively) were much lower than those in corn-based fermentation. In metabolic pathway of *C. acetobutylicum*, butyrate formation and re-assimilation reaction constitute a closed-loop (butyrate loop) at butyryl-CoA node (Figure 1). The butyrate loop not only competes with butanol synthesis route for carbon resource, but also relates to acetone formation. Therefore, high butanol/acetone ratio in cassava-based fermentation was also attributed to weakened metabolic strength of butyrate closed-loop. It could be concluded that higher NADH generation rate and lower metabolic flux in butyrate closed-loop worked jointly, leading to the high butanol/acetone ratio feature in fermentation with cassava-based substrate.

The alteration of redox balance to promote NADH generation has been reported in many studies using corn or glucose as substrate. The methods of provision of artificial electron carriers such as neutral red [26],[18] and methyl viologen [27]-[29], ORP regulation [30], and inhibition of hydrogenase by spraying carbon monoxide [31],[32], all of them have effect on enhancing butanol/acetone ratio. Among these approaches, adding neutral red seems to be the most appropriate one of low-cost and easy to operate. In the previous works, it has been demonstrated that adding neutral red could acquire a 63% increment in butanol/acetone ratio with corn-based substrate [26],[18]. Therefore, neutral red was added into cassava broth at 60 h when butanol production rate was in a relatively high level, in order to further enhance butanol/acetone ratio. The results indicated that final butanol concentration could slightly increase, due to the enhancements in r_{NADH} and *adhE* transcriptional level after neutral red addition. However, the metabolic fluxes of organic acid formation/re-assimilation pathway and *ctfAB* transcription level were also enhanced after supplementing neutral red (Figures 5 and 6), which was actually beneficial for acetone formation. The simultaneous enhancement of both butanol and acetone synthesis route led to an unchanged butanol/acetone ratio in cassava-based fermentation with neutral red addition. It was speculated that under the reductive compound-enriched environment using cassava-based substrate, reductive power NADH might have been excessively produced/consumed leading to a burden on cellular metabolism [33]. To match up relatively low fluxes of the acid loops with enhanced NADH regeneration in an appropriate way will be the key issue in obtaining further high butanol/acetone ratio while maintaining comparably high butanol productivity.

In addition to strength of the reductive power, many other approaches have been adopted to acquire high butanol/acetone ratio. Among these efforts, some modifications to metabolic pathways were obtained good effect on increasing butanol/acetone ratio. Harris et al. used metabolic engineering tools to restrain *buk* (encoding butyrate kinase) expression and resulted in a significant increase in butanol/acetone ratio (3.8) [34]. Jang et al. further strengthened the butanol synthesis pathway by up-regulating *adhE* expression, on the basis of the weakened acid formation pathway, which achieved a rather high butanol/acetone ratio (8.8) [35]. However, some other attempts to regulate pathways had totally different results. Lehmann et al. sought to construct a *pta/ptb* double knockout strain of *C. acetobutylicum* but failed to find any positive clones [36]. Tummala attempted to downregulate the expression of *ctfAB*, but observed high organic acid accumulation in broth with low solvent production [37]. It could be found that all these efforts to extend the butanol/acetone ratio were involved in the modifications of acid formation pathways. It is known that acetate and butyrate formation pathways are the major energy substance (ATP) production route in *C. acetobutylicum*. As obligate anaerobes, clostridia are rather inefficient in energy production. So an irreversible change on the major energy substance (ATP) production route would inevitably bring adverse impact on cell growth and solvent production. In this study, a 64% increment in butanol/acetone ratio was obtained by changing corn to a cheaper feedstock cassava, without irreversible damages to cell or cost increase (the additional amount of yeast extract was rather less, and the cost of cassava-added yeast extract was even much lower than that of corn). Moreover, most of the mechanism analysis researches in the past were based on concise culture environment by using a defined medium. Therefore, this mechanism research based on biomass substrate was necessary for achieving high butanol/acetone ratio under industrial ABE fermentation condition.

In addition, it was noteworthy that why higher NADH generation rate and lower metabolic flux of butyrate closed-loop appeared in cassava-based culture environment not corn? What special chemical compositions of cassava resulted in these phenomena? Aiming to illustrate the above questions, another research is now carried out by us. Currently, it is discovered that when carbon/nitrogen ratio (C/N ratio) in the substrate is increased from 46.7 to 186.7 mol/mol, acid formation is visibly restrained in the solventogenic phase, leading to a 28% reduction in acetone production, but no adverse impact on butanol production. Moreover, H_2 generation dropped off rapidly, associated with C/N ratio increase, which implied that high C/N ratio could limit the electron directing to H_2 synthesis and contribute to enhance NADH production. These results indicated that higher C/N ratio contained in cassava (118.8 mol/mol) compared to corn (21.9 mol/mol) may be one of the factors leading to higher butanol/acetone ratio. Besides, there are still other potential factors needed to be further confirmed. The toxic compounds contained in cassava, which may have caused the restraint of some enzyme activities such as CoA-transfer or butyrate kinase, is a notable aspect and needs to be explored deeply.

CONCLUSIONS

A higher butanol/acetone ratio of approximately 2.9:1 was observed in extractive fermentation on cassava-based substrate, which had a 64% increment compared to that on corn-based substrate. The results from metabolic flux and transcriptional analysis indicated that the weakened metabolic fluxes in organic acids (especially butyrate) formation/ re-assimilation pathways, as well as the enhancement of NADH generation, contributed to this higher butanol/acetone ratio feature in fermentation on cassava. Moreover, neutral red added in cassava broth could not further increase butanol/acetone ratio, which demonstrated that a further higher butanol/acetone ratio could be realized only when NADH regeneration is enhanced and the metabolic fluxes in organic acid formation/reutilization routes are controlled at suitably low levels simultaneously.

AUTHORS' CONTRIBUTIONS

XL carried out the fermentation experiment, performed the statistical analysis, and drafted the manuscript. Z-GL carried out the fermentation experiment. Z-PS conceived of the study and participated in its design and coordination and helped to draft the manuscript. All authors read and approved the final manuscript.

ACKNOWLEDGEMENTS

The study was supported by the National Natural Science Foundation Program (#20976072) and Major State Basic Research Development

Program (#2007CB714303) of China. The authors also appreciated the assistance from Mr. Kevin Ding in refining the English of the manuscript.

REFERENCES

1. Jones DT, Woods DR (1986) Acetone-butanol fermentation revisited. Microbiol Rev 50(4):484-524

2. Lee SY, Park JH, Jang SH, Nielsen LK, Kim J, Jung KS (2008) Fermentative butanol production by clostridia. Biotechnol Bioeng 101(2):209-228

3. Pfromm PH, Amanor-Boadu V, Nelson R, Vadlani P, Madl R (2010) Bio-butanol vs. bio-ethanol: a technical and economic assessment for corn and switchgrass fermented by yeast or *Clostridium acetobutylicum*. Biomass Bioenergy 34(4):515-524

4. Liu Z, Ying Y, Li F, Ma C, Xu P (2010) Butanol production by *Clostridium beijerinckii* ATCC 55025 from wheat bran. J Ind Microbiol Biotechnol 37(5):495-501

5. Qureshi N, Ezeji TC, Ebener J, Dien BS, Cotta MA, Blaschek HP (2008) Butanol production by *Clostridium beijerinckii*. Part I: use of acid and enzyme hydrolyzed corn fiber. Bioresour Technol 99(13):5915-5922

6. Qureshi N, Saha BC, Dien B, Hector RE, Cotta MA (2010) Production of butanol (a biofuel) from agricultural residues: Part I - use of barley straw hydrolysate. Biomass Bioenergy 34(4):559-565

7. Qureshi N, Saha BC, Hector RE, Dien B, Hughes S, Liu S, Iten L, Bowman MJ, Sarath G, Cotta MA (2010) Production of butanol (a biofuel) from agricultural residues: Part II - use of corn stover and switchgrass hydrolysates. Biomass Bioenergy 34(4):566-571

8. Lu C, Zhao J, Yang ST, Wei D (2012) Fed-batch fermentation for n-butanol production from cassava bagasse hydrolysate in a fibrous bed bioreactor with continuous gas stripping. Bioresour Technol 104:380-387

9. Li X, Li Z, Zheng J, Shi Z, Li L (2012) Yeast extract promotes phase shift of bio-butanol fermentation by *Clostridium acetobutylicum* ATCC824 using cassava as substrate. Bioresour Technol 125:43-51

10. Harris LM, Blank L, Desai RP, Welker NE, Papoutsakis ET (2001) Fermentation characterization and flux analysis of recombinant strains of *Clostridium acetobutylicum* with an inactivated solR gene. J Ind Microbiol Biotechnol 27(5):322-328

11. Desai RP, Harris LM, Welker NE, Papoutsakis ET (1999) Metabolic flux analysis elucidates the importance of the acid-formation pathways in regulating solvent production by*Clostridium acetobutylicum*. Metab Eng 1(3):206-213

12. Hoenicke D, Janssen H, Grimmler C, Ehrenreich A, Luetke-Eversloh T (2012) Global transcriptional changes of Clostridium acetobutylicum cultures with increased butanol:acetone ratios. New Biotechnol 29(4):485-493

13. Bankar SB, Survase SA, Singhal RS, Granstrom T (2012) Continuous two stage acetone-butanol-ethanol fermentation with integrated solvent removal using *Clostridium acetobutylicum* B 5313. Bioresour Technol 106:110-116

14. Dhamole PB, Wang Z, Liu Y, Wang B, Feng H (2012) Extractive fermentation with non-ionic surfactants to enhance butanol production. Biomass Bioenergy 40:112-119

15. Roffler SR, Blanch HW, Wilke CR (1987) In-situ recovery of butanol during fermentation. Bioprocess Biosyst Eng 2(1):1-12

16. Evans PJ, Wang HY (1988) Enhancement of butanol formation by *Clostridium acetobutylicum*in the presence of decanol-oleyl alcohol mixed extractants. Appl Environ Microbiol 54(7):1662-1667

17. Zhang L, Yang Y, Shi Z (2008) Performance optimization of property-improved biodiesel manufacturing process coupled with butanol extractive fermentation. Chin J Biotechnol 24(11):1943-1948

18. Li Z, Li X, Zheng J, Zhang S, Shi Z (2011) Butanol extractive fermentation to simultaneously produce "properties improved" biodiesel and butanol in a water and energy-saving operation way. J Biobased Mater Bioenergy 5(3):312-318

19. Papoutsakis ET (1984) Equations and calculations for fermentations of butyric acid bacteria. Biotechnol Bioeng 26(2):174-187

20. Meyer CL, Papoutsakis ET (1989) Increased levels of ATP and NADH are associated with increased solvent production in

continuous cultures of *Clostridium acetobutylicum*. Appl Microbiol Biotechnol 30(5):450-459

21. Takiguchi N, Shimizu H, Shioya S (1997) An on-line physiological state recognition system for the lysine fermentation process based on a metabolic reaction model. Biotechnol Bioeng 55(1):170-181

22. Gu Y, Hu S, Chen J, Shao L, He H, Yang Y, Yang S, Jiang W (2009) Ammonium acetate enhances solvent production by *Clostridium acetobutylicum* EA 2018 using cassava as a fermentation medium. J Ind Microbiol Biotechnol 36(9):1225-1232

23. Ishizaki A, Snibai H, Hirose Y (1974) Basic aspects of electrode potential change in submerged fermentation. Agric Biol Chem 38:2399-2405

24. Demuez M, Cournac L, Guerrini O, Soucaille P, Girbal L (2007) Complete activity profile of*Clostridium acetobutylicum* FeFe-hydrogenase and kinetic parameters for endogenous redox partners. FEMS Microbiol Lett 275(1):113-121

25. Gheshlaghi R, Scharer JM, Moo-Young M, Chou CP (2009) Metabolic pathways of clostridia for producing butanol. Biotechnol Adv 27(6):764-781

26. Girbal L, Vasconcelos I, Sant-Amans S, Soucaille P (1995) How neutral red modified carbon and electron flow in *Clostridium acetobutylicum* grown in chemostat culture. FEMS Microbiol Rev 16:151-162

27. Peguin S, Goma G, Delorme P, Soucaille P (1994) Metabolic flexibility of *Clostridium acetobutylicum* in response to methyl viologen addition. Appl Microbiol Biotechnol 42:611-616

28. Rao G, Mutharasan R (1987) Altered electron flow in continuous cultures of *Clostridium acetobutylicum* induced by viologen dyes. Appl Environ Microbiol 53(6):1232-1235

29. Rao G, Mutharasan R (1986) Alcohol production by *Clostridium acetobutylicum* induced by methyl viologen. Biotechnol Lett 8(12):893-896

30. Wang S, Zhu Y, Zhang Y, Li Y (2012) Controlling the oxidoreduction potential of the culture of *Clostridium acetobutylicum* leads to an earlier initiation of solventogenesis, thus increasing solvent productivity. Appl Microbiol Biotechnol 93(3):1021-1030

31. Kim BH, Bellows P, Datta R, Zeikus JG (1984) Control of carbon and electron flow in*Clostridium acetobutylicum* fermentations: utilization of carbon monoxide to inhibit hydrogen production and to enhance butanol yields. Appl Environ Microbiol 48(4):764-770

32. Meyer CL, Roose JW, Papoutsakis ET (1986) Carbon monoxide gasing leads to alcohol production and butyrate uptake without acetone formation in continuous cultures of*Clostridium acetobutylicum*. Appl Microbiol Biotechnol 24:159-167

33. Savinell JM, Palsson BO (1992) Network analysis of intermediary metabolism using linear optimization. I. Development of mathematical formalism. J Theor Biol 154(4):421-454

34. Harris LM, Desai RP, Welker NE, Papoutsakis ET (2000) Characterization of recombinant strains of the *Clostridium acetobutylicum* butyrate kinase inactivation mutant: need for new phenomenological models for solventogenesis and butanol inhibition? Biotechnol Bioeng 67(1):1-11

35. Jang YS, Lee JY, Lee J, Park JH, Im JA, Eom MH, Lee J, Lee SH, Song H, Cho JH, Seung DY, Lee SY (2012) Enhanced butanol production obtained by reinforcing the direct butanol-forming route in *Clostridium acetobutylicum*. Mbio 3(5):e00314-12

36. Lehmann D, Radomski N, Luetke-Eversloh T (2012) New insights into the butyric acid metabolism of *Clostridium acetobutylicum*. Appl Microbiol Biotechnol 96(5):1325-1339

37. Tummala SB, Junne SG, Papoutsakis ET (2003) Antisense RNA downregulation of coenzyme A transferase combined with alcohol-aldehyde dehydrogenase overexpression leads to predominantly alcohologenic *Clostridium acetobutylicum* fermentations. J Bacteriol 185(12):3644-3653

The Glycerol Biorefinery: a Purpose for Brazilian Biodiesel Production

Emerson Léo Schultz, Daniela Tatiane de Souza, and Mônica Caramez Triches Damaso

Embrapa Agroenergy, Parque Estação Biológica, PqEB s/n – Av.W3 Norte (final), Brasília, 70770-901, DF, Brazil

ABSTRACT

According to estimates from the International Energy Agency, global energy consumption will increase by at least one third, between 2010 and 2035. The additional power required will be provided not only by fossil sources but also by renewables. While the world energy matrix is supplied only by 13.2% from renewable sources, Brazil has different scenery with renewables accounting for 42.4% of the energy matrix. This work aimed to evaluate the potential use of oleaginous in biorefineries considering the produced quantity, prices, and costs of raw materials and products. Considering the availability of these raw materials, the results showed significant opportunities that can be exploited in Brazil, within the biorefinery concept. Soybean oil is the main raw material for biodiesel production in Brazil, although there are many other vegetable oils with potential for this purpose. Related to the production costs, the soybean biodiesel has higher costs than diesel. Then, this biofuel is only produced due to Brazilian regulatory rules and public subsidies. In order to become this production favorable in the market environment, it is essential to aggregate value to all byproducts and residues generated along the biodiesel production chain. Glycerin is a byproduct of biodiesel that could be used, in a glycerol biorefinery concept, as raw material for the production of value-added products through chemical, biochemical, or thermochemical routes.

INTRODUCTION

Nowadays, most of the energy consumed in the world comes from non-renewable sources, such as petroleum, coal, and natural gas. The world energy matrix uses only 13.2% from renewable sources, while the Brazilian scenario differs from this context using 42.4% of these sources, according to 2012 statistical data. They are represented mainly by sugarcane biomass (15.4%), hydraulic and electric (13.8%), natural gas (11.5%), and firewood and charcoal (9.1%). The participation of renewable sources in the energetic matrix is among the highest in the world, although there has been a slight reduction (1.8%) from 2011 to 2012, due to lower supply of hydropower and ethanol [1].

Factors such as Brazil's vast territory and favorable weather conditions expand the possibility of using water and biomass for energy production. Among the biomasses, the main are sugarcane, oilseeds, and lignocellulosic materials (e. g., crop residues and wood chips). Notwithstanding, many Brazilian industrial sectors that use biomass as the main raw material have envisioned its use not only to produce energy, but in a wider perspective using the concept of biorefinery. For these sectors, the application of this concept is a great opportunity for growth and market expansion.

A biorefinery is a facility that integrates biomass conversion processes and equipment to produce fuels, power, and chemicals [2]. A biorefinery is not a completely new concept. Many of the traditional biomass converting technologies such as sugar, starch, and pulp and paper industries use aspects connected with this approach. However, several economic and environmental drivers such as global warming, security of supply, energy supply, high energy costs, and agricultural policies have also directed those industries to further evolve their operations into biorefineries [3].

The combinations involving raw materials, conversion/technology processes, and final products are almost unlimited in the biorefinery concept. The final decision about what product will be prioritized in a biorefinery (biofuels, chemicals, or bioenergy) will depend on the availability of raw materials, technological knowledge, public policies, regulations, and market dynamics.

Considering the availability of raw materials and favorable conditions of price and costs, Brazil has significant opportunities that can be exploited in the biorefinery concept. Thus, this work aimed to present the overview of the Brazilian biodiesel production and the perspective of glycerin use in the biorefinery context.

REVIEW

Raw Materials for Biodiesel Production

In 2004, due to the necessity to expand the supply of energy from biomass, the Brazilian Government introduced biodiesel in the national energy matrix, through the National Program for Production and Use of Biodiesel [4]. In 2005, was established optional addition of 2% biodiesel to petroleum diesel (named B2), which became mandatory from January 2008. This percentage was increased successively by the National Energy Council (CNPE) up to 5% (B5) in January 2010.

In Brazil, currently, there are 64 biodiesel plants authorized for operation by the Brazilian National Agency of Petroleum, Natural Gas and Biofuels (ANP), with a total production capacity of 7 million m³/year [5]. In 2012, more than 2.5 million m³ of biodiesel were produced [6], which corresponds to 36% of the current installed capacity. This production placed Brazil in the third position in the world ranking of biodiesel producers.

The main commercial route used in the biodiesel production is the transesterification process of vegetable oils and fats with alcohols, in the presence of a basic catalyst. This reaction results in three molecules of fatty acid monoalkyl esters, which compose the biodiesel, and one molecule of glycerol or glycerin (byproduct) [7].

In Brazil, there are more than 200 species of plants like fruits and grains that can produce oil, with different potentialities and natural adaptations to environmental conditions, which can be used to produce biofuels or other value-added products [8], but only some of them have been used for biodiesel production. Data on production of oilseeds and fat used for this purpose are shown in Table 1.

Table 1: Brazilian production of oilseeds and bovine fat in 2011 [[9]-[11]]

Product	Harvested area (ha)	Production (tonnes)	Oil production (kg ha−1)	Main producing state
Soybean	23,968,663	74,815,447	540	Mato Grosso
Cottonseed	1,405,135	5,070,717	360	Mato Grosso
Palm (bunch)	109,080	1,301,192	4,000	Pará
Bovine fat	Na	750,000	Na	São Paulo
Peanut	106,679	311,459	800	São Paulo
Castor beans (berry)	208,476	120,166	705	Bahia
Sunflower	62,535	77,932	630	Mato Grosso
Canolaa	42,400	52,000	500	Rio Grande do Sul

Na, not applicable. aProduction refers to the 2011/2012 season and not the calendar year.

Schultz et al.

Schultz et al. Chemical and Biological Technologies in Agriculture 2014 1:7, doi:10.1186/s40538-014-0007-z

Figure 1 presents the contribution of each oil source used for Brazilian biodiesel production in April 2013. The soybean is the main raw material used. Most of the national soybean production (about 44%) is exported in the form of grain, 7% is stocked, and 49% is processed domestically for obtaining 79% of soybean meal and 21% of oil. Seventy-seven percent of this oil is used for food and biodiesel production [12]. In April 2013, the soybean oil used for biofuel production represents more than 2.0 million m^3 [13]. The soybean is the main raw material for biodiesel production as a consequence of some important issues, such as technological domain of the crop, scale of production, and the success in establishing the productive chain across the country [8].

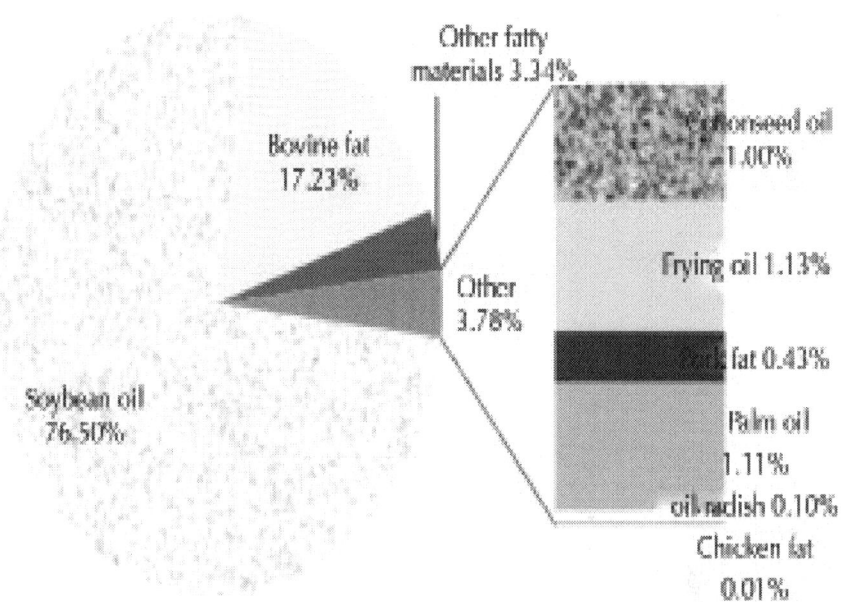

Figure 1: Raw materials used in the production of biodiesel (B100) in Brazil: April 2013 [[13]].

Figure 2 displays the growth in the use of the main sources of oils for biodiesel production. Although the role of soybean oil in the biodiesel production is really the most important, the use of fat in this matrix has increased substantially, probably, because it is a residue and its cost is lower than oilseeds. The main barrier for the expansion of fat in biodiesel matrix is the high number of saturated fat acids present in this material.

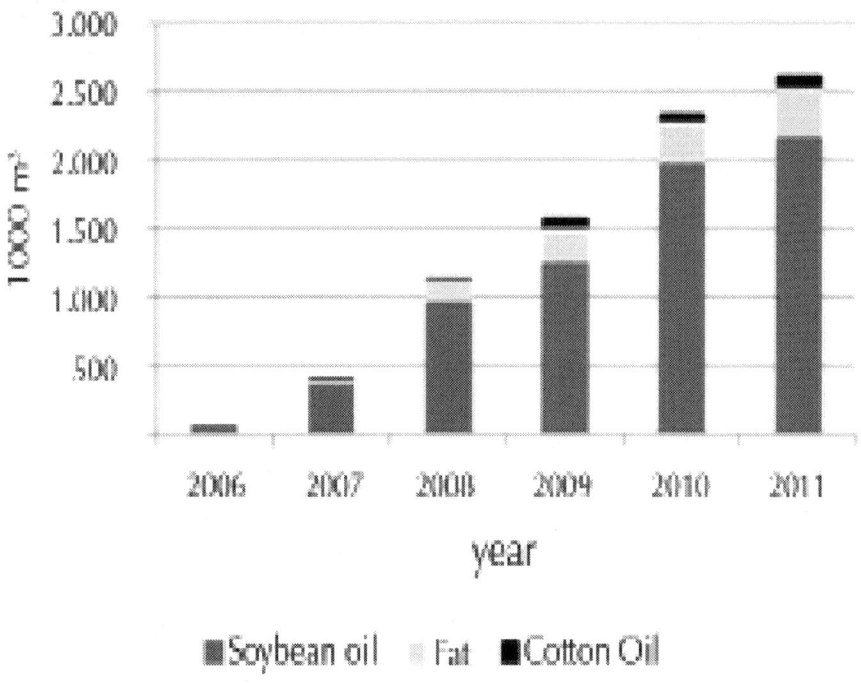

Figure 2: The main oil sources used in the Brazilian biodiesel production [[14]].

There is an expectation that the regulatory framework will be increased to B10 by the Brazilian Government. If this actually occurs, it will be urgent to increase the raw material supply through the diversification in the production of oilseed crops in various regions of the country.

Prices and Costs of Soybean Biodiesel

The price of biodiesel depends greatly on the price of soybean oil in Brazil. In Figure 3, it can be seen that there is a strong correlation between the nominal prices of biodiesel, practiced in the auctions of the National Petroleum Agency (ANP), and the prices of soybean oil. The correlation found between these two variables in the period of January 2011 to November 2013 was 0.73. The National Petroleum Agency uses it largely in the price of soybean oil for pricing in auctions.

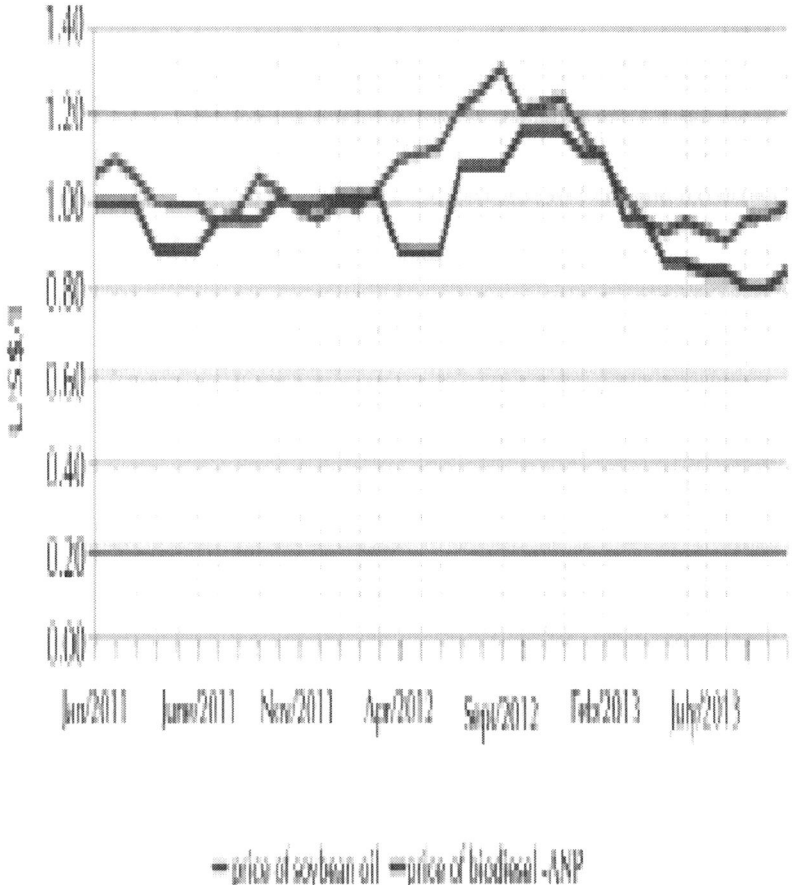

Figure 3: Evolution of biodiesel prices in the ANP auctions and the prices of crude soybean oil (quotation = R$2.30/US$). Source: author's calculations based on [15] and [16].

With the launch of the National Biodiesel Program (PNPB) establishing the mandatory blending and the scheduled increase in the proportion of biodiesel in fossil diesel, there was a consolidation of the biodiesel market in Brazil. The growth of investments in the structure of production of biodiesel allowed for an increase to 5% blending with 3 years in advance. Achieving this goal has been attributed to the ability of the industry to meet the market needs, as well as the will of the Brazilian Government in the growth of the use of this energy source [17].

The price of biodiesel in the country is set at auctions and it is fixed for a period of 3 months. On the other hand, the soybean oil is a commodity and its price is set by the market. As with other agricultural commodities, after 2002, there was a rise in the price of soybean oil, a fact that reduced substantially the margins of biodiesel processing plants. More recently, there was a fall in the price of soybean oil, reaching US$0.99/L in November 2013 [16]. This fall has been attributed to the high amount of soybean harvested in Brazil, which has enhanced the availability of raw material for biofuel production. Moreover, the decline is also attributed to the worldwide availability of vegetable oils and excess capacity of the biodiesel facilities in the country (next to 60%) [18].

Thus, there was a drop in sales prices of B100 at auctions in 2013. The average price of biodiesel in the ANP auctions in 2013 was US$0.90/L, well below the price established in October 2012 (US$1.16/L). Even with this drop in current prices and reduced profit margins, other factors allow the plants to continue to produce biodiesel and expand its production capacity as the growth potential in underexplored markets is high [17].

Table 2 shows the prices and production costs of diesel and biodiesel. In 2013, the average selling price of diesel in the distributor was US$0.88/L [15]. As transport costs, profit margins, and taxes account for 70% of this value [19], it is estimated that US$0.26/L corresponds to the production cost of diesel.

Table 2: Comparison of prices and costs of diesel and soybean biodiesel [[15],[19],[20]]

Description	Price/cost as of 2013 (US$/L)
Price of diesel in the refinery [15]	0.69
Price of diesel in the distribution [15]	0.89
Production cost of diesel (estimated in this paper)	0.26
Average price of biodiesel [15]	0.90
Cost of biodiesel production (alkaline transesterification with a residual oil from the pre-treatment step of oil) [20]	0.63
Cost of biodiesel production [19]	0.65

Cost of biodiesel production (conventional alkaline transesterification) [20]	1.25

Schultz et al.

Schultz et al. Chemical and Biological Technologies in Agriculture 2014 1:7, doi:10.1186/s40538-014-0007-z

Different authors estimate different cost of biodiesel production in Brazil [19], [20]. These costs undoubtedly vary according to the production process as well as the raw material. The cost of soybean biodiesel can reach values between US$0.63/L and US$1.25/L, excluding taxes, shipping, and producer profit margin. This value is well above the values of the production cost of diesel (US$0.26/L).

The high cost of biodiesel production by alkaline transesterification is bound mainly to the high cost of soybean oil, which corresponds to 77% of total manufacturing costs [20]. Soybean is considered the main source of raw material with enough availability to meet the demand for biodiesel in the country.

Challenges and Opportunities within the Concept of Glycerol Biorefinery

It is estimated that 90 m^3 of biodiesel produced by transesterification generates approximately 10 m^3 of glycerin. Figure 4 shows the increase of glycerin production which follows the biodiesel production growth in the period from 2005 to 2011. If the use of B10 is implemented in Brazil, 5.56 million m^3 biodiesel would be produced, corresponding to approximately 600,000 m^3 of glycerin, which is much higher than the national demand. Furthermore, glycerin has a low market value.

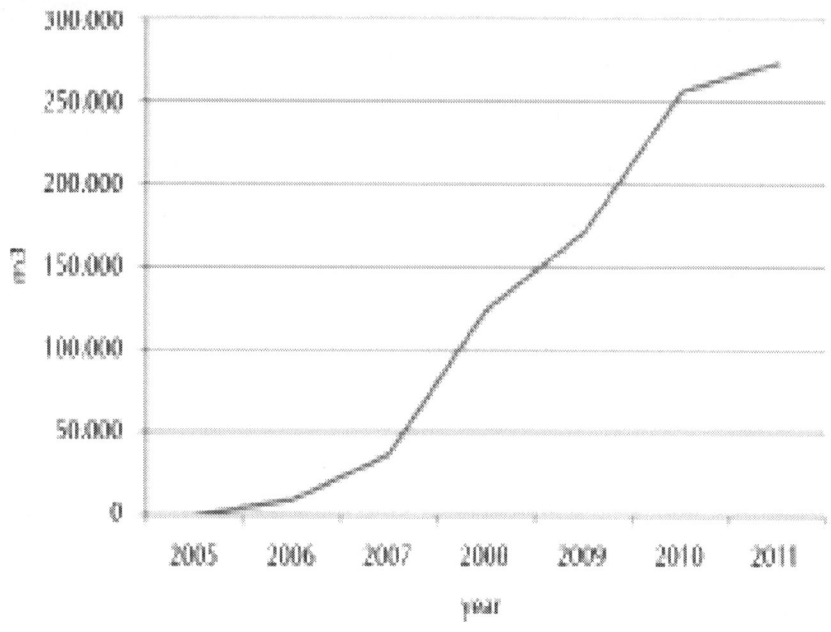

Figure 4: Glycerin generated in the production of biodiesel (B100) in Brazil: 2005 to 2011 [[14]].

The increasing production of glycerin is not only a Brazilian reality; other countries such as the USA [21] and Germany also have growth in the production of crude glycerin, which is directly related to the increase in the production of biodiesel.

In addition to the excessive amount of glycerin produced, another important issue is the quality of this material. As the main production process of biodiesel is a chemical route using basic catalysts, crude glycerin contains diverse impurities such as alcohol, traces of the catalyst, esters, and salts that hinder its direct application in several industrial processes. Nowadays, the main industrial sectors that consume glycerin worldwide are cosmetics, soaps, and pharmaceuticals [22],[23] that require glycerin with certain degree of purity. The purification of crude glycerin obtained from biodiesel production to transform it into analytical grade is not economically favorable for sectors that work with lower value-added products [24].

The crude glycerin produced in Brazil is mostly used to generate heat by burning it in industrial furnaces and boilers, as in biofuel production, besides potteries and steel companies. Also, Brazilian glycerin has been exported to China [25].

Brazil exported 16,400 tonnes of glycerin with revenue of US$5.4 million, equivalent to US$0.30/kg of glycerin [26]. However, exports of glycerin decreased more than 45% in March of 2013 [27], resulting in a large accumulation of this byproduct in biodiesel plants.

Since there is a global trend of continuing increase in the consumption of biodiesel, it is necessary to search for different technological solutions for crude glycerin in order to add value to the productive chain of biodiesel within the biorefinery concept. Similarly to the sugarcane biorefinery, it has been proposed for glycerol biorefinery [28]. Glycerin may be used as raw material for the production of value-added products through chemical, biochemical, or thermochemical routes, as shown in Figure 5[23],[24],[28]-[32].

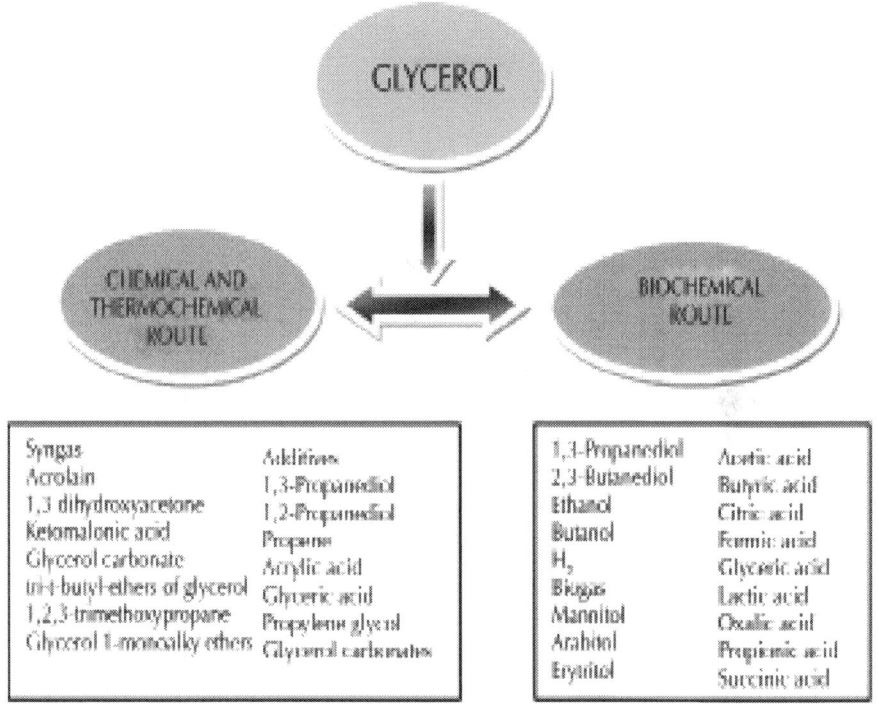

Figure 5: Products obtained from glycerin produced in biorefinery [[23],[24],[28]-[32]].

Currently, another interesting solution can be to apply crude glycerin from vegetable oil biodiesel production as raw material for animal feed. Some reported data indicate that the use of 10% crude glycerin as a pig and catfish complement feed maintains their growth with normal performance [30].Ethanol, hydrogen, and syngas could be used directly as fuels or as intermediates in chemical synthesis. Chemical compounds as polyols and organic acids are very important as building blocks. They are used in industrial sectors such as cosmetic, pharmaceutical, automobile, and chemical industries [33].

Glyceric acid may be used in chemical and pharmaceutical industries as a building block and for the production of polymers and surfactants. Lactic acid can be processed to make acrylic acid or 1,2-propanediol used in polyester resins and polyurethane. Succinic acid is largely used for manufacturing health-related products and as building block to produce precursors that are converted into green solvents, pharmaceutical products, and biodegradable plastics. Polyols are used in food, pharmaceutical, and medical industries. They are used to improve the nutritional profile of food products due to their low caloric content, low insulin-mediated response, and non-cariogenicity. Polyols and their derivatives also have other industrial applications, including the production of polyurethanes, plastifying agents, resins, surfactants, and intermediates for producing hydrocarbons [23],[24],[28]-[33].

The chemical routes used to transform glycerin include selective oxidation, etherification, dehydration, hydrogenolysis, and reforming [23],[34]-[37]. In the selective oxidation, the oxidation of primary hydroxyl groups yields glyceric acid and tartronic acid, both of which are commercially useful compounds. Dihydroxyacetone (DHA), an important fine chemical, is obtained from the oxidation of the secondary hydroxyl groups, while ketomalonic (or mesoxalic) acid results when all three hydroxyl groups are oxidized [34],[38].

Alkyl ethers can be synthesized by the reaction of glycerol with alkenes and preferentially with isobutylene in the presence of an acid catalyst. Ether derivatives of glycerol reduce emissions of particulate matter, hydrocarbons, carbon monoxide, and aldehydes, when incorporated in standard diesel fuel containing 30% to 40% of aromatic compounds [34],[39].

Dehydration of glycerol may follow two pathways. The dehydration of the secondary hydroxyl group yields 3-hydroxypropanal, which undergoes a dehydration step, leading to the formation of acrolein. On the other hand, the dehydration of the primary hydroxyl group results in 1-hydroxyacetone, also known as acetol [23].

Hydrogenolysis is a catalytic chemical reaction that breaks a chemical bond in an organic molecule with the simultaneous addition of a hydrogen atom to result in molecular fragments. Propylene glycol (1,2-propanediol (1,2-PD)) and 1,3-propanediol (1,3-PD) could be obtained through the hydrogenolysis of glycerol in the presence of metallic catalysts and hydrogen [23],[36].

One promising way is to use glycerin to produce synthesis gas via steam reforming [40]. Overall, glycerin steam reforming can be represented by the following reaction [37]:

$$C_3H_8O_3 + H_2O \rightarrow 3CO_2 + 7H_2$$

(1)

Another process used to produce synthesis gas from glycerol is the aqueous phase reforming (APR), using aqueous solutions of glycerol and a catalyst [41]. Synthesis gas can be used to produce hydrocarbons (through Fischer-Tropsch synthesis), methanol, hydrogen, isobutene, and isobutane [42]. BioMCN has been producing methanol from synthesis gas obtained from glycerin since 2010 in Delfzijl (Netherlands) [35].

One process widely known for the valorization of glycerin is the synthesis of epichlorohydrin. This chemical is employed in the production of epoxy resins. Advanced Biochemical (Thailand) Co., a subsidiary of Vinythai (shareholders are Solvay and PTT), is producing 100,000 t/year of epichlorohydrin in Thailand (Map Ta Phut) using Solvay technology (Epicerol). A new plant of Vinythai in China (Taixing) should become operational in the second half of 2014, also with capacity of 100,000 t/year and using the Epicerol technology. In this process, glycerol reacts with HCl, yielding a mixture of 1,2-dichloro-3-propanol and 1,3-dichloro-2-propanol, which is treated with NaOH resulting in epichlorohydrin. The Dow Chemical Co. and Spolchemie (Czech Republic) have also developed processes to convert glycerin to epichlorohydrin [23],[34],[35],[43]-[45].

Figure 6 shows some products and their intermediates obtained by biotechnological pathway of glycerol (glycerin) conversion.

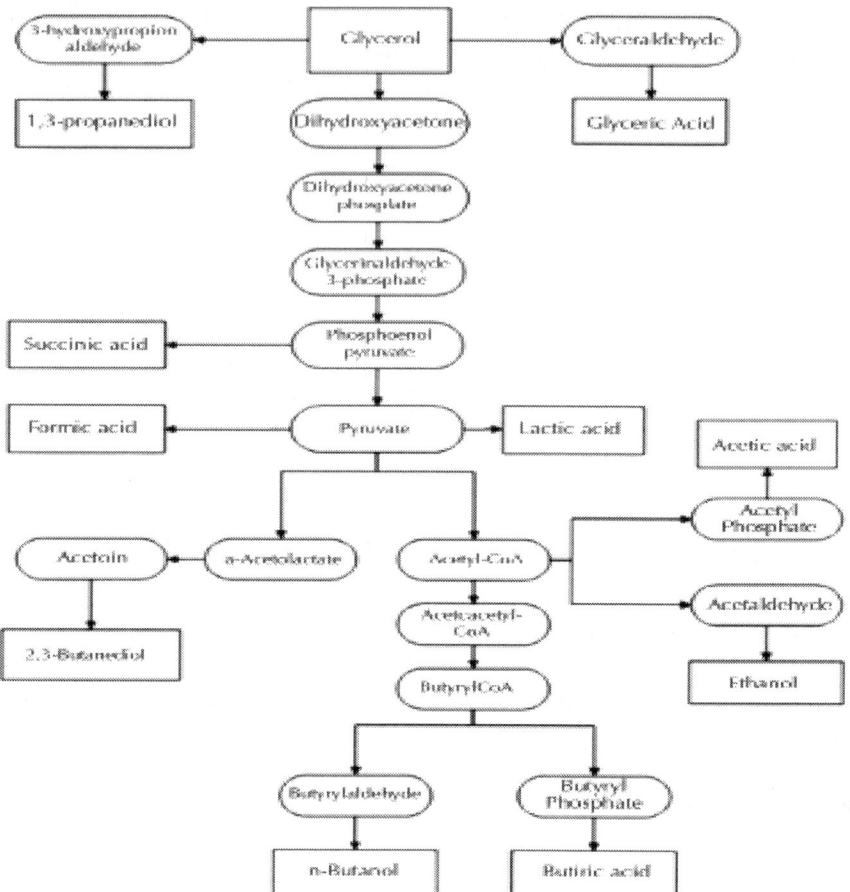

Figure 6: Products obtained from glycerol bioconversion metabolism [[28],[29],[46],[47]]. Products and glycerol are enclosed in boxes and the main intermediates are enclosed in circles.

The route of biotechnological valorization of glycerol can follow three different pathways. Yeast and filamentous fungi convert glycerol to polyols and organic acids by aerobic process. Enterobacteriaceae produces butanol and 2,3-butanediol using anaerobic way. Additionally,*Escherichia coli* and other bacteria produce lactic and glyceric acids and dihydroacetone under microaerobic conditions. Succinic acid can be produced by aerobic and microaerobic pathways, while 1,3-propanediol and ethanol can be obtained by anaerobic and microaerobic conditions [28],[29],[46],[47].

Notwithstanding, there are some aggregate-value bioproducts that can be obtained from glycerin by biotechnological route. Although there are several works in the literature and patents, commercial initiatives are still rare. Some examples of industrial scale production are described below; however, there is no information about industrial initiatives in Brazil.

The chemical companies BASF and Cargill and industrial biotechnology company Novozymes are developing bio-based technologies to produce acrylic acid from renewable feedstocks. Currently, acrylic acid is produced by the oxidation of propylene derived from the refining of crude oil. Using biotechnology route, these companies have worked to develop microorganisms that can efficiently convert renewable feedstock, among them are glycerol, into 3-hydroxypropionic acid, which is one possible chemical precursor to acrylic acid [48]. The companies have carried out this precursor production in pilot scale, but they intend to increase for commercial scale still in 2014.

1,3-propanediol has showed to be a scientific and technological interest product. An Asia (METabolic EXplorer) and European initiatives (Propanergy and Glyfinery Projects) have been carried out to construct pilot and industrial plants to obtain this product from glycerin raw material. In all cases, the biotechnological route and the biorefinery concept are applied, including the production of biogas and other aggregated-value bioproducts. To develop the European projects, partnerships between private companies and universities were established [49]-[51].

Posada et al. [52] evaluated the conversion of raw glycerol to nine added-value products obtained by chemical (synthesis gas, acrolein, and 1,2-propanediol) or biochemical (ethanol, 1,3-propanediol, lactic acid, succinic acid, propionic acid, and poly-3-hydroxybutyrate) routes. The results showed that not only quality requirements were successfully reached but that all the processes were profitable. More income can be earned from 1,3-propanediol and 1,2-propanediol production, while less income would be obtained from hydrogen (synthesis gas) and succinic acid.

According to recent analysis of technical papers, the most studied chemical products obtained from glycerin are 1,3-propanediol, hydrogen, and propylene glycol. On the other hand, propylene glycol, acrolein, and dichlorohydrin were the most cited chemical products in patents. Additionally, the main chemicals obtained from operating plants are propylene glycol and epichlorohydrin [53].

In this context, if glycerin is used in Brazil or worldwide as raw material for the production of fuels and value-added products in the glycerol biorefinery concept, it could increase balance of trade through the sale of these products, and it can prevent the accumulation of glycerin in companies when exports are unfavorable.

CONCLUSIONS

Brazil stands out on using biomass as an energy source. In addition, many raw materials have great potential for use in biorefineries. Among these raw materials are oilseeds such as soybeans, besides forest biomass, sugar cane and other biomass. Considering the availability of these raw materials and favorable conditions of price and costs, the results showed significant opportunities that can be exploited in Brazil in the biorefinery concept.

Soybean oil is the main raw material used in the production of biodiesel in Brazil. Despite increased use of biomass of soybean in the Brazilian energy matrix, petroleum products remain the most economically viable. The cost of production of soybean biodiesel in Brazil (US$1.25/L) is well above the values of the production cost of diesel (US$0.26/L). This is certainly a challenging factor that encourages the consolidation of biorefineries in Brazil. Besides biodiesel from soybean oil, the application of the biorefinery concept aimed the production of a wide range of products. The production of chemicals and biofuels from glycerin biodiesel byproduct is one of the main challenges within the concept of biorefinery. Some products as polyols, organic acids, alcohols, syngas, and biogas could be obtained from this industry that could be called 'glycerol biorefinery'.

Both biochemical and chemical route, including thermochemical processes, are essential and complementary in a biorefinery. It is not yet possible to define the best technologies and products for oleaginous chains that are under development, but at the present time, all of them must be considered. The possible combinations of raw materials, conversion/technology processes, and final products are practically unlimited in the biorefinery concept.

AUTHORS' CONTRIBUTIONS

MCTD and ELS wrote the Abstract and the 'Introduction' and 'Raw materials for biodiesel production' sections. DTS wrote the section 'Prices and costs of soybean biodiesel'. All authors participated in the writing of 'Challenges and opportunities within the concept of glycerol biorefinery' and the 'Conclusions' sections. All authors read and approved the final manuscript.

ACKNOWLEDGEMENTS

We would like to thank Maria Goreti Braga for the graphical abstract design and José Manuel Cabral de Sousa Dias for the critical text review. This work was supported by grants from the Brazilian Agricultural Research Corporation (Embrapa) and Brazilian National Council for Scientific and Technological Development (CNPq).

REFERENCES

1. Year 2102 / Empresa de Pesquisa Energética EPE, Rio de Janeiro
2. King D, Inderwildi OR, Williams A, Hagan A (2010) The future of industrial biorefineries World Economic Forum, Geneva
3. Publications IEA Bioenergy Task 42 Biorefinery. [http:/ / www. iea-bioenergy.task42-biorefiner ies.com/ upload_mm/ 5/ 8/ 2/ a47bd297-2ace-44d0-92bbcb7cc02f75de _Brochure_Totaal_ definitief_webs.pd f]. Accessed 18 Feb 2014
4. Pousa GPAG, Santos ALF, Suarez PAZ (2007) History and policy of biodiesel in Brazil. Energy Policy. 35:5393–5398
5. ANP Monthly Bulletin of Biodiesel. [http://www.anp.gov. br/?dw=65297]. Accessed 10 Feb 2014
6. ECDFAOAgriculturalOutlook.[http://stats.oecd.org/viewhtml. aspx?QueryId=36356&vh=0000 &vf=0&l&il=blank&lang=em] Accessed 19 Feb 2014
7. Geris R, Santos NAC, Amaral BA, Maia IDS, Castro VD, Carvalho JRM (2007) Biodiesel de soja: reação de transesterificação para aulas práticas de química orgânica. Quim Nova. 30:1369–1373

8.   Laviola BG, Alves AA (2011) Matérias-primas oleaginosas para biorrefinarias. In: Vaz S Jr (ed) Biorrefinarias: cenários e perspectivas, Embrapa Agroenergia, Brazil

9.   IBGE Produção Agrícola Municipal. [http://www.sidra.ibge.gov.br/bda/tabela/listabl.asp?c=1612&z=p&o=18] Accessed 19 Feb 2014

10. Conab Acompanhamento da Safra Brasileira. [http://www.conab.gov.br/ OlalaCMS/ uploads/ arquivos/ 12_01_10_10_53_02_ boletim_graos_4o_ levantamento.pdf]. Accessed 19 Feb 2014

11. Garcia S [http:/ / pecuaria.ruralbr.com.br/ noticia/ 2012/ 02/ sebo-bovino-e-alternativa-viavel-pa ra-producao-debiodiesel-3670782.htm l] Accessed 19 Feb 2014

12. APROSOJA. [http://www.aprosoja.com.br]. Accessed 19 Feb 2014

13. ANP Monthly bulletin of biodiesel. National Agency of Petroleum, Natural Gas and Biofuels, Rio de Janeiro

14. ANP. [http://www.anp.gov.br/?dw=61225]. Accessed 19 Feb 2014

15. ANP Preços nominais do biodiesel. [http://www.anp.gov.br/?dw=67024]. Accessed 19 Feb 2014

16. ABIOVE Estatística mensal do complexo soja. [http://www.abiove.org.br/site/index.php?page=estatistica&area=NC0yLTE] =. Accessed 19 Feb 2014

17. NETO S A R Biodiesel. [http:/ / conab.gov.br/ OlalaCMS/ uploads/ arquivos/ 11_04_11_08_45_49_conjuntura_biodis el_ marco_2011.pdf] Accessed 19 Feb 2014

18. Caetano M Preço do biodiesel é o menor desde o início da mistura obrigatória. [http:/ / www.valor.com.br/ agro/ 3327022/ preco-do-biodiesel-e-o-menor-desde- o-inicio-da-mistura-obrigatoria] Valor Econômico, 04 /11/2013. . Accessed 19 Feb 2014

19. Cavalett O (2008) Análise do ciclo de vida da soja. In: PhD Thesis Department of Food Engineering, Universidade Estadual de Campinas, SP, Brazil

20. Viabilidade técnico-econômica da produção de biodiesel via rota alcalina e supercrítica baseadas em óleo residual. In: Masters Thesis Department of Chemical Engineering, Universidade Federal do Ceará, Fortaleza, CE, Brazil

21. Jonhson DT, Taconi KA (2007) The glycerin glut: options for the value-added conversion of crude glycerol resulting from biodiesel production. Environ Prog. 26:338–348

22. Beatriz A, Araújo YJK, Lima DP (2011) Glicerol: um breve histórico e aplicação em sínteses estereosseletivas. Quim Nova. 34:306–319

23. Mota CJA, Silva CXA, Gonçalves VLC (2009) Gliceroquímica: novos produtos e processos a partir da glicerina de produção de biodiesel. Quim Nova. 32:639–648

24. Li Y, Reeder R (2011) Turning crude glycerin into polyurethane foam and biopolyols. In: Agriculture and Natural Resources Fact Sheet. The Ohio State University, Wooster, OH, USA. [http:// ohioline. [http://ohioline.osu.edu/aex-fact/pdf/0654.pdf] osu. edu/aex-fact/pdf/0654.pdf]

25. Parente E, Jr Resíduos bem-vindos. [http://revistapesquisa.fapesp. br/2012/06/14/residuos-bem-vindos]

26. BiodieselBR. [http:/ / www.biodieselbr.com/ noticias/ usinas/ glicerina/ exportacoes-glicerina-us-54-mi-feve reiro-120313. htm]

27. (2013) BiodieselBR. [http:/ / www.biodieselbr.com/ noticias/ usinas/ glicerina/ exportacoes-glicerina-caem-45-marco -150413.htm]

28. Choi WJ (2008) Glycerol-based biorefinery for fuels and chemicals. Recent Pat Biotechnol. 2:173–180

29. Almeida JR, Fávaro LC, Quirino BF (2012) Biodiesel biorefinery: opportunities and challenges for microbial production of fuels and chemicals from glycerol waste. Biotechnol Biofuels. 5:1–48

30. Leoneti AB, Aragão-Leoneti V, Valle S, Oliveira WB (2012) Alternatives for the use of unrefined glycerol. Renew Energy. 45:138–145

31. Fan XO, Burton R, Zhou Y (2010) Glycerol - byproduct of biodiesel production as a source for fuels and chemicals. Open Fuels Energy Sci J. 3:17–22

32. Pachauri N, He B (2006) Value-added utilization of crude glycerol from biodiesel production: a survey of current research activities.

In: Proceedings of the ASABE Annual International Meeting American Society of Agricultural and Biological Engineers, Portland

33. Bozell JJ, Petersenb GR (2010) Technology development for the production of biobased products from biorefinery carbohydrates—the US Department of Energy's "Top 10" revisited. Green Chem. 12:525–728

34. Pagliaro M, Ciriminna R, Kimura H, Rossi M, Della Pina C (2007) From glycerol to value-added product. Angew Chem Int Ed. 46:4434–4440

35. Katryniok B, Paul S, Dumeignil F (2013) Recent developments in the field of catalytic dehydration of glycerol to acrolein. ACS Catal. 3:1819–1834

36. Zhou CHC, Beltramini JN, Fan YX, Lu GQM (2008) Chemoselective catalytic conversion of glycerol as biorenewable source to valuable commodity chemicals. Chem Soc Rev. 37:527–549

37. Adhikari S, Fernando SD, Haryanto A (2008) Hydrogen production from glycerin by steam reforming over nickel catalysts. Renew Energy. 33:1097–1100

38. Gil S, Cuenca N, Romero A, Valverde JL, Sáchez-Silva L (2014) Optimization of the synthesis procedure of microparticles containing gold for the selective oxidation of glycerol. Appl Catal A. 2014(472):11–20

39. Arco Chemical Technology, Kesling LP, Jr HS, Karas LJ, Liotta FJ Diesel fuel. Patent. Int. Cl. C10L 1/00. US5308365. 31/08/1993, 03/05/1994

40. Avasthi KS, Reddy RN, Patel S (2013) Challenges in the production of hydrogen from glycerol—a biodiesel byproduct via steam reforming process. Procedia Eng. 51:423–429

41. Alonso DM, Bond JQ, Dumesic JA (2010) Catalytic conversion of biomass to biofuels. Green Chem. 12:1493–1513

42. Spath PL, Dayton DC Preliminary screening—technical and economic assessment of synthesis gas to fuels and chemicals with emphasis on the potential for biomass-derived syngas. [http://www.nrel.gov/docs/fy04osti/34929.pdf]

43. Solvay to build bio-based epichlorohydrin plant in China to serve largest market in the world. [http://www.solvay.com/en/media/press_releases/20120611-epicerol-china.html]

44. Vinythai company profile. [http://www.solvayplastics.com/sites/solvayplastics/EN/vinyls/Vinythai/Pages/CompanyProfile.aspx]

45. Bell BM, Briggs JR, Campbell RM, Chambers SM, Gaarenstroom PD, Hippler JG, Hook BD, Kearns H, Kenney JM, Kruper WJ, Schreck DJ, Theriault CN, Wolfe CP (2008) Glycerin as renewable feedstock for epichlorohydrin production. The GTE process. Clean 36:657–661

46. Abad S, Turon X (2012) Valorization of biodiesel derived glycerol as a carbon source to obtain added-value metabolites: focus on polyunsaturated fatty acids. Biotechnol Adv. 30:733–741

47. Biebl H, Menzel K, Zeng AP, Deckwer WD (1999) Microbial production of 1,3-propanediol. Appl Microbiol Biotechnol. 52:289–297

48. BASF, Cargill and Novozymes target commercial bio-based acrylic acid process. [http://www.basf.com/group/pressrelease/P-12-363]

49. METABOLIC EXPLORER: METabolic EXplorer announces start-up of industrial-scale pilot. [http://www.euroinvestor.com/news/2009/12/14/metabolic-explorer-metabolic-explorer-announces-start-up-of-industrial-scale-pilot/10786665]

50. Periodic report summary 2 - PROPANERGY (integrated bioconversion of glycerine into value-added products and biogas at pilot plant scale). [http://cordis.europa.eu/result/rcn/53625_en.html]

51. GLYFINERY: sustainable and integrated production of liquid biofuels, green chemicals and bioenergy from glycerol in biorefineries. [https://www.ifeu.de/landwirtschaft/pdf/glyfinery_theparliamentmagazine_march2010.pdf]

52. Posada JA, Rincón LE, Cardona CA (2012) Design and analysis of biorefineries based on raw glycerol: addressing the glycerol problem. Bioresour Technol. 111:282–293

53. Freitas ZS (2013) Glicerina como matéria-prima para a indústria química: avaliação dos esforços de pesquisa e das iniciativas comerciais. In: Master Thesis Escola de Química, Universidade Federal do Rio de Janeiro, Rio de Janeiro, RJ, Brazil

Chapter 7

Heavy Metal Pollution in Ancient Nara, Japan, During the Eighth Century

Hodaka Kawahata[1,2], Shusuke Yamashita[2], Kyoko Yamaoka[3], Takashi Okai[3], Gen Shimoda[3], and Noboru Imai[3]

[1]Atmosphere and Ocean Research Institute, The University of Tokyo, Kashiwanoha 5-1-5, Kashiwa, Chiba 277-8564, Japan

[2]Department of Earth and Planetary Science, Faculty of Science, The University of Tokyo, Hongo 7-3-1, Bunkyo-ku, Tokyo 113-0033, Japan

[3]Geological Survey of Japan, National Institute of Advanced Industrial Science and Technology (AIST), Higashi 1-1-1, Tsukuba, Ibaraki 305-8567, Japan

ABSTRACT

We quantitatively investigated the eighth century heavy metal pollution in Heijo-kyo (Ancient Nara), the first large, international city of Japan. In this metropolis, mercury, copper, and lead levels in soil were increased by urban activity and by the construction of the

Great Buddha statue, Nara Daibutsu. Mercury and copper pollution associated with the construction of the statue was particularly high in the immediate vicinity of the statue, but markedly lower in the wider city environment. We therefore reject the hypothesis that extensive mercury pollution associated with the construction of the Nara Daibutsu made it necessary to abandon Ancient Nara, even though severe lead pollution was detected at several sites. The isotopic composition of the lead indicated that it originated mainly from the Naganobori mine in Yamaguchi, which was a major source of the copper for the Nara Daibutsu.

BACKGROUND

Human civilizations are physical spheres of human activity on earth. This human activity is typically associated with pollution, which results in environments becoming dangerous to live in, or in which chemical substances or energy may become limited (Matsui 2007). Although pollution increased dramatically during the industrial revolution, the material cycle was also affected by ancient human activity, especially mining and smelting. For example, the rate of atmospheric Pb deposition in the Jura Mountains of Switzerland has been affected appreciably by human activity over the last 2,100 years, with rates of deposition during this period being at least 10 times that of the original natural flux (Shotyk et al. 1998). The establishment of large cities in ancient times would lead to chemical pollution, mainly due to their large populations and the commencement of the metal-working industries. However, the extent of these ancient episodes of pollution has not yet been quantified.

Heijo-kyo (Ancient Nara) was Japan's second capital and the first large and international city (from 710 to 784 A.D.). Although Fujiwara-kyo, the layout of which is based on the grid of a Chinese city, is officially considered to be Japan's first capital, it was a small city that only served as the capital for 6 years after construction was completed. Heijo-kyo is a registered UNESCO World Heritage site (http://whc.unesco.org/en/list/870) (Figure 1). It had a maximum population of 50,000 to 200,000, making it half the size of Constantinople (Istanbul) at that time (Japan Broadcasting Corporation 2012), and a high population density of 2,000 to 8,000 people km^{-2}. The city covered approximately 25 km^2

and had an irregular rectangular shape modeled after Chang'an, the capital city of Tang Dynasty China. Numerous foreign visitors traveled to Ancient Nara where they introduced their cultures. These visitors included traders and Buddhist monks from China, Korea, India, and Vietnam. Buddhism prospered and was particularly influential in politics where it was adopted by the government and propagated by foreign priests. As a result, the Tenpyo Culture was developed, a lowery culture strongly connected to Buddhism (UNESCO World Heritage site; Hall 1993). A number of great temples and religious structures were established during this time, one of which was Todaiji Temple, which is famous for its very large Buddha statue known in Japanese as the Nara Daibutsu. The Great Buddha of Nara is the world's largest gilt bronze statue of Buddha. The statue, which is approximately 15 m tall and is seated in a cross-legged position (Figure 2), was built in response to repeated outbreaks of smallpox in Japan in the 730 s A.D. (Hall 1993). The statue was constructed in the following stages: (1) A rough framework was constructed from wood and other materials and then covered with micaceous clay, from which the features of the statue were sculpted. (2) After the sculpted clay surface had dried, a mixture of sand and clay was laid over the clay form, baked, and removed in sections to serve as the outer mold. (3) After removal of the outer mold, the clay form was pared down so that it could serve as the inner mold. The completion of these three stages took almost 2 years. (4) The outer mold was reassembled around the inner mold and (5) smelted bronze was poured into the space between the two molds. Because the statue was so high, the last two stages were repeated eight times over the course of 4 years to complete the casting, starting from the bottom of the statue and working toward the top. (6) After the outer mold was removed, the surface of the statue was gilded by the amalgamation method. To make the amalgam, Au was dissolved in Hg at a ratio of about 1:5. After this fluid was applied to the statue, the surface was broiled to evaporate the Hg, leaving a uniform layer of Au on the surface. Emperor Shomu issued an imperial decree to build the great statue of Buddha, intending to resolve the national crisis through Buddha's divine protection (Hall 1993). The hall of the Buddha was also the largest ancient wooden structure in the world, although the 'Metropol Parasol' in Seville, Spain, is currently the largest such structure in the world (Moore 2011).

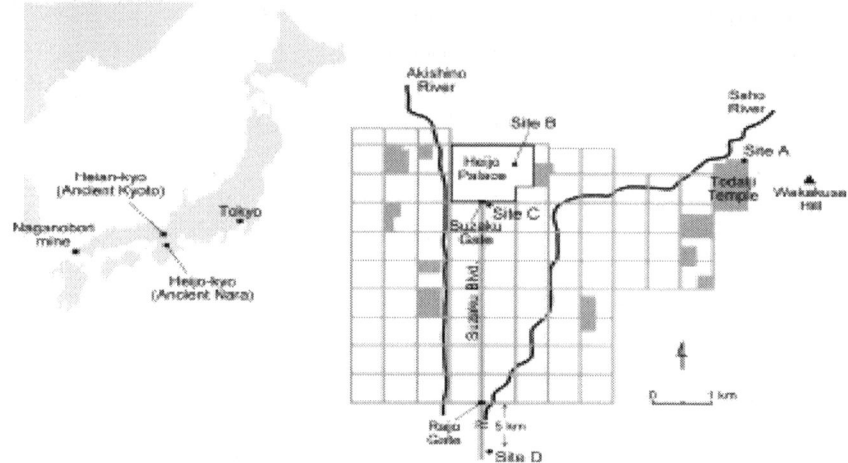

Figure 1: Location of Ancient Nara in Japan, with inset showing soil sampling sites and other relevant locations in and around the city.The streets of Ancient Nara had ditches on both sides. These ditches functioned as drains or sewers, carrying polluted water into the main rivers. People may have indiscriminately disposed of waste materials in such ditches, occasionally even human skulls.

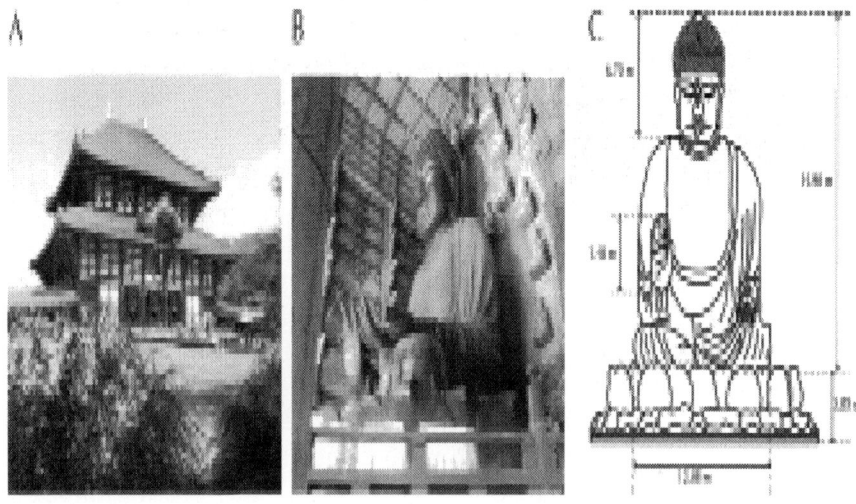

Figure 2: Photographs and schematic representation of the Nara Daibutsu. (A) Current Great Buddha Hall (Daibutsuden), which is 57.5 m wide, 50.5

m deep, and 49.1 m tall. It was rebuilt in 1691 A.D. due to fire. The original hall was even larger and had a width of 86 m. (B, C) Great Buddha. Processes involved in the production of the Buddha were as follows: More than 2 million people were reportedly involved in this project, even though the total population of Japan was 5 million at that time (Kito 2000).

It has been considered that Ancient Nara may have been polluted by urban activity, particularly by activities associated with the construction of the Nara Daibutsu, which was made of 499,000 kg of Cu and 8,500 kg of Sn and gilded with 146 kg of Au and 820 kg of Hg (Konishi 2002). The statue was gilded by the amalgamation method, in which an amalgam of Hg and Au was applied to the surface and then heated to evaporate the Hg. Shirasuga (2002) hypothesized that the Hg that was released into the environment around Ancient Nara due to the construction of the Nara Daibutsu caused people in the area to suffer significantly from an unidentified 'strange disease' and that the government was forced to relocate to a new capital for fear of an evil curse or divine punishment. He also proposed that careless disposal of waste materials from the copper refining process was primarily responsible for a treeless, grass-covered hill located to the east of Todaiji Temple. Although these theories have little supporting historical or scientific evidence, they have nevertheless persisted in popular culture and are often cited in Japanese television programs on the Nara Daibutsu. Another explanation for the relocation of the capital city may have been because Emperor Kammu wished to relocate the capital to an area with better transportation routes, and/or to escape the power of the Buddhist clergy (Hall 1993).

Urban activity is known to pollute the environment on both local and regional scales, and in a city as large as Ancient Nara, people would have relied on the natural environment to dispose of their sewage and other waste. They may have buried their waste in large holes or in ditches, practices which may have led to the accumulation of waste and pollutants. In this study, we provide quantitative data for the presence of toxic heavy metals, such as Hg, Pb, Cu, and Fe, in ancient soil samples and remains collected from archeological sites in Ancient Nara. Furthermore, we evaluate the effect of urban activity on ancient metal pollution and discuss the environmental effects associated with the construction of the Nara Daibutsu.

CASE DESCRIPTION

Background and Climate of Modern Nara

The archeological site of Ancient Nara is located within the modern city of Nara, which is surrounded by mountains on three sides and is located at the northern end of the Nara Basin in the Kansai region of Japan (Figure 1). The dominant soils consist of terrace deposits, derived from metamorphic mudstone and biotite granite from the Saho River. Metal pollution has not been reported in this area in recent times as mining for metals is no longer conducted in the region (Ozaki 2000).

Nara has a mild climate, with mean temperatures ranging between 3.8°C (January) and 26.6°C (August). The mean annual rainfall is 1,333 mm and the direction of the prevailing winds is southeast in summer and northwest in winter (Nara Meteorological Observatory 2014).

Sampling Sites

Ancient soil samples were carefully collected from four historic sites in and around Ancient Nara (Figure 1) under the supervision of archeologists (Table 1). The age of each sample was determined based on any artifacts that were present, from writings on mokkan (long, thin, narrow pieces of wood strung together that were used to write on in ancient times), and from other archeological evidence. The sampled soil sediments had not been disturbed and heavy metal contamination by subsequent overprinting is not suspected because Ancient Nara became an agricultural area after the capital moved to Kyoto, and the soil sediments remained deeply buried for more than 1,000 years (Hall 1993).

Table 1: Soil metal concentrations and Pb isotopic ratios for samples from Ancient Nara and Naganobori mine

Site	Sample	Age	Sample	Hg ppb	Cu ppm	Pb ppm	Fe ppm	Mn ppm	Co ppm	Ni ppm	Zn ppm	208Pb/206Pb	207Pb/206Pb
A	Dark gray gravel containing copper slag	Middle eighth century	1-A	92	370	100	34,000	170	21	14	60	2.094	0.850
A	Dark gray gravel containing copper slag	Middle eighth century	1-B	66	57	32	36,000	240	23	14	52		
A	Dark gray clay	Twelfth century	2	298	62	18	18,000	140	12	N.D.	26		
A	Light orange-colored sand	Sixteenth century	3	255	56	210	56,000	170	27	29	120		
B	Brownish soil containing wood debris	Late eighth century	100	229	18	90	24,000	550	17	11	47		
B	Dark gray soil containing wood debris	Late eighth century	106	207	8.6	130	14,000	95	9	3.3	30	2.093	0.851
B	Dark gray soil containing wood debris	Late eighth century	118	133	12	16	15,000	98	7.8	3.5	28		
C	River bed	Just before the eighth century	203	45	5.6	200	5,800	92	5.2	4.1	23	2.093	0.851
C	Soil preparation for Heijo-kyo	Early to middle eighth century	204	86	11	1,100	30,000	630	17	15	38	2.093	0.851

Soil preparation for Heijo-kyo	Early to middle eighth century	205	78	4.8	470	12,000	520	9.4	9.4	27			C
Muddy soil around pillar hole	Early to middle eighth century	206	92										C
Muddy soil around pillar hole	Early to middle eighth century	207	59	5.7	380	10,000	140	4.9	5.7	20			C
Muddy soil burying the well-frame	Eighth century	201	107	4.6	500	8,200	150	6.3	5.1	25	2.093	0.851	C
Muddy soil burying the well-frame	Eighth century	202	103	5.7	15	7,400	140	8.8	6.7	50			C
Muddy soil in dump site around smithy	Eighth century	208	157	18	1,200	57,000	2,300	32	22	39	2.093	0.851	C
Muddy soil in dump site around smithy	Early ninth century	209	155										C
Muddy soil in paddy field	Medieval period	210	212	9.3	18	24,000	200	7.2	6.5	26			C
Muddy soil	Age of 1870 to 1960	211	521	21	230	11,000	190	4.8	11	50			C
Muddy soil in pillar hole	Early to middle eighth century	212	141										C

Group	Sample description	Period	Sample no.	Value									
C	Muddy soil in pillar hole	Early to middle eighth century	213	165									0.851
C	Muddy soil in pillar hole	Early to middle eighth century	214	118									2.093
C	Muddy soil in pillar hole	Early to middle eighth century	215	104									
C	Muddy soil in pillar hole	Early to middle eighth century	216	137									
C	Muddy soil in pillar hole	Early to middle eighth century	217	106									
C	Muddy soil in paddy field	Medieval period	218	185									
D	Brownish sandy soil	Late eighth century	31	276	13	740	61,000	1,200	20	10	110		
D	Dark gray silty soil	Middle eighth century	34	298	14	880	55,000	730	24	13	140		
D	Dark gray sandy soil	Very early eighth century	38	39	2.3	13	21,000	320	14	N.D.	35		

Naganobori mine									
Slag	300,000	360	180,000	1,800	500	3.9	8,000	2.092	0.846
Slag	110,000	62	210,000	1,100	81	1.2	1,400	2.092	0.847
Chrysocolla	620,000	21	270	8.1	N.D.	N.D.	4,100	2.091	0.847

Kawahata et al.

Kawahata et al. Progress in Earth and Planetary Science 2014 1:15, doi: 10.1186/2197-4284-1-15

Site A is located 20 m north of a section of the hall that housed the Nara Daibutsu in Todaiji Temple in approximately 750 A.D. (Figure 3). Three kinds of samples were taken from a pit that was excavated for the construction of a drain pipe in October 2011. The samples were classified based on a description of the cross section of this section by the Buried Cultural Property Investigation Center (Figure 3; personal communication with Mr. M. Kanekata and Mr. T. Yasui; Board of Education of Nara City 2011). Sampling was supervised by Mr. Kanekata and Mr. Yasui (Board of Education of Nara City 2011). Sample 1 was dated to the middle of the Nara period (around 750 A.D.) on the basis of the presence of many copper particles originating from the bronze blast furnace used in the construction of the Nara Daibutsu (Ishino 2004). Therefore, sample 1 is assumed to represent direct metal contamination (Figure 3). Samples 2 and 3 were dated to the 12th and 16th centuries, respectively, because they contained debris and soils which corresponded to fires in 1180 and 1567 A.D. that destroyed the wooden structures housing the statue.

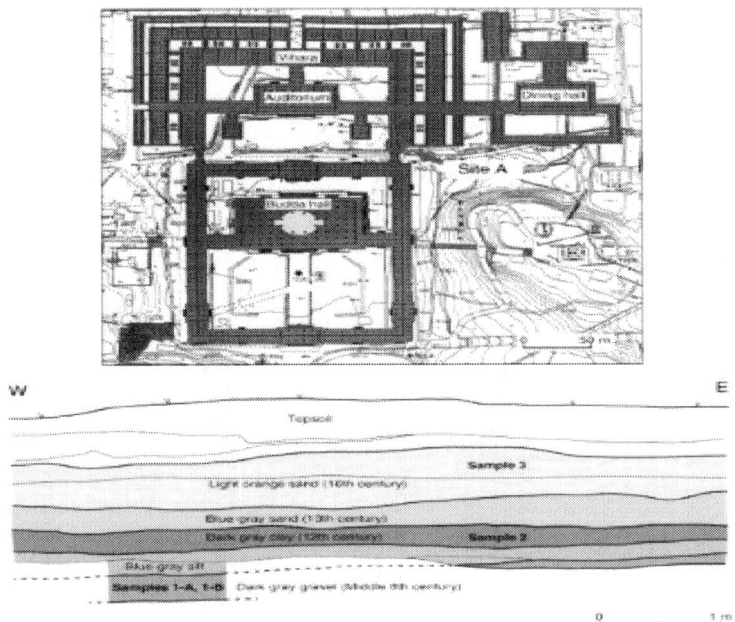

Figure 3: Detailed soil map at Site A and sketch showing exact location and stratigraphic position of samples. Source: Board of Education of Nara City 2011.

Site B was situated inside the central palace, the home of the imperial family and the location of the central governmental offices. Currently, excavations and investigations of this area are performed under the auspices of the Nara National Research Institute for Cultural Properties. These samples were collected from a dump in this area in order to estimate the pollution associated with garbage at the palace in the very center of Ancient Nara. The artifacts present in these samples were dated from the late eighth century (Nara National Research Institute for Cultural Properties 2010, 2012a).

Site C was located in front of Suzaku-mon, the south main gate of the palace. Eighteen samples were collected in order to evaluate the levels of pollution in the early stages of urban activity: the soil layers here corresponded to the foundation of Ancient Nara in the early eighth century, and iron blacksmiths are considered to have operated around the corner in the early eighth century (Nara National Research Institute for Cultural Properties 2012a, [b]; personal communication, Dr. Jinno). Samples 203, 204, 205, 206, and 207 were from the early eighth century. Sample 210 dated from a rice paddy in the 9th to 11th centuries.

Site D was located approximately 5 km south of the center of Ancient Nara (Archaeological Institute of Kashihara 2012a, [b]). Ten soil samples were collected from a ditch along Suzaku Boulevard, which was an extension of the central avenue, supervised by Dr. Kinoshita and other archeologists working for the Archaeological Institute of Kashihara in Nara Prefecture. Three of these samples were selected for the analysis of sewage and to determine the level of background pollution in the ancient city. The samples consisted of muddy soils from the early to late eighth century (Figure 4) (Archaeological Institute of Kashihara 2012a, [b]).

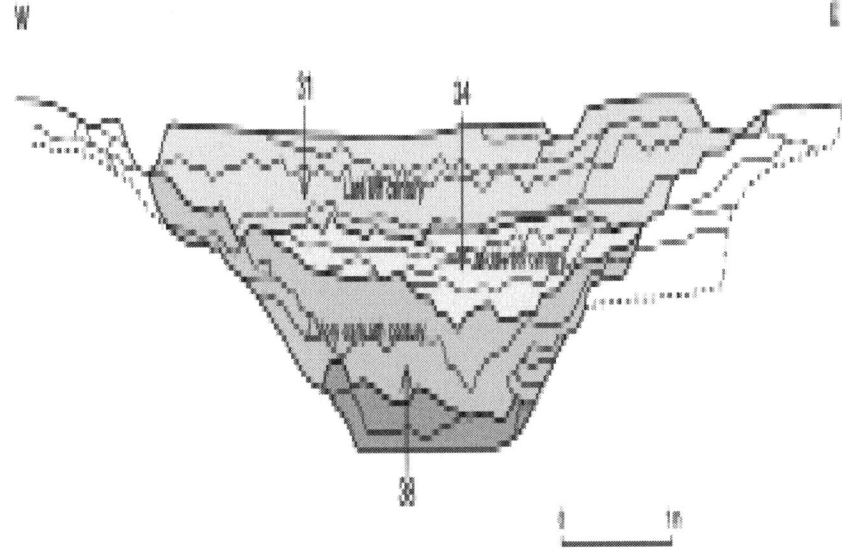

Figure 4: Sketch showing exact location and stratigraphic position of samples at Site D. Source: Archaeological Institute of Kashihara, 2012a, [b].

The Naganobori mine in Yamaguchi Prefecture in western Honshu is a skarn-type deposit that produced mainly Cu and Pb. The ore is a contact deposit between Paleozoic Akiyoshi-dai limestone and granite porphyry. The mine had been used since the late seventh century and is known to have supplied almost all of the Cu used in the construction of the Nara Daibutsu (Mitochou Compilation Committee 2004). The Naganobori mine was called Nara-nobori mine in seventh century because 'Nara-nobori' means 'Cu was transported up to the capital city of Nara'.

Analytical Procedures

Elemental Analysis

Each sample was crushed into fine powder and separated into subsamples for elemental analysis. A 200 mg aliquot of each dry bulk sample was digested in an ultrapure mixture of HF and HNO_3, and any organic residue was removed by filtration. The solutions were then diluted to 100 ml with ultrapure (Milli-Q) water and then subjected to analysis at the Geological Survey of Japan. Fe, Co, and Ni were analyzed by inductively coupled plasma atomic emission spectrometry (ICP-AES: Seiko Instruments SPS7800, Chiba, Japan). Cu, Pb, Mn, and Zn were analyzed using inductively coupled plasma mass spectrometry (ICP-MS: Agilent 4500 UCO-MS, Tokyo, Japan). Hg was analyzed using vapor atomic absorption spectrometry (AAS: Japan Instruments Mercury/MA-2000, Tokyo, Japan). A reference rock standard (JSO-1, Geological Survey of Japan) was used to calibrate samples as well as standard solutions prepared from pure elemental standard solutions (Wako Pure Chemical Industry Ltd., Osaka, Japan). Analytical error was estimated to be less than 10% for each elemental analysis (Kawahata et al. 2006), which is considered to be smaller than the variation attributed to ubiquity in soil.

Pb Isotope Analysis

The digested samples were dried and dissolved in 0.5 M HBr. Two slag samples and chrysocolla from the Naganobori mine were also prepared after decomposition by $HF-HNO_3$. Pb was then purified using 0.5 ml anion exchange resin (Bio-Rad AG 1-X8). Other elements were eluted using 3.0 ml of 0.5 M HBr, and the Pb fraction was collected using 4.0 ml of 6 M HCl. The procedural Pb blank was 50 pg during chemical separation, which was negligible compared with the sample values of >1,200 ng Pb. Pb isotopic ratios were determined at the Geological Survey of Japan using MC-ICP-MS (NEPTUNE multicollector ICP-MS, Thermo Finnigan, Germany) with a Tl doping technique. Samples were diluted to approximately 200 ppb. Pb based upon previous determinations. In addition to four Pb isotopes (^{204}Pb, ^{206}Pb, ^{207}Pb and ^{208}Pb), ^{202}Hg was also measured to subtract the ^{204}Hg signal from ^{204}Pb. A sample-standard bracketing method was used to correct for instrumental mass bias, using a Pb standard solution prepared from a reference standard (SRM 981, National Institute of Standards and Technology, Gaithersburg, MD, USA). For comparison, a Pb solution prepared from another reference standard (SRM 982, National Institute of Standards and Technology, USA) was also analyzed. The mean $^{206}Pb/^{204}Pb$, $^{207}Pb/^{204}Pb$, and $^{208}Pb/^{204}Pb$ values of NIST SRM981 ($n = 34$) were 16.9315 ± 0.0025, 15.4847 ± 0.0031, and 36.6772 ± 0.0079, respectively (mean ± 2 standard deviations).

DISCUSSION AND EVALUATION

The results of metal and Pb isotope analyses for each of the soil samples are given in Table 1. Hg content ranged from 39 to 521 ppb, with a mean of 160 ppb. The two lowest values, 39 and 45 ppb, might reflect the background Hg level in Ancient Nara because the corresponding samples (samples 203 and 38) date to the late seventh century or early eighth century (Figure 5). Samples 204, 205, 206, and 207, which also had relatively low Hg contents (59 to 92 ppb), date to the early and middle eighth century. Samples 1-A and 1-B had low Hg contents (66 and 92 ppb, respectively); these samples were obtained from heaps of soil produced during the construction of the Nara Daibutsu and, therefore, date from just before the mercury-gold gilding of the statue was undertaken. Conversely, samples 2, 3, and 210, which are younger than that of the ninth century, had Hg contents of at least 200 ppb. The maximum value of 521 ppb was observed in sample 211, which dates from between 1870 and 1960 and is, thus, probably due to modern contamination. Large amounts of Hg were released into the environment in the 19th and 20th centuries due to the operation of coal-fired power plants and the use of Hg in electric lights, batteries, electrodes for sodium hydroxide production, and wood preservation (Japan Oil, Gas and Metals National Corporation 2010).

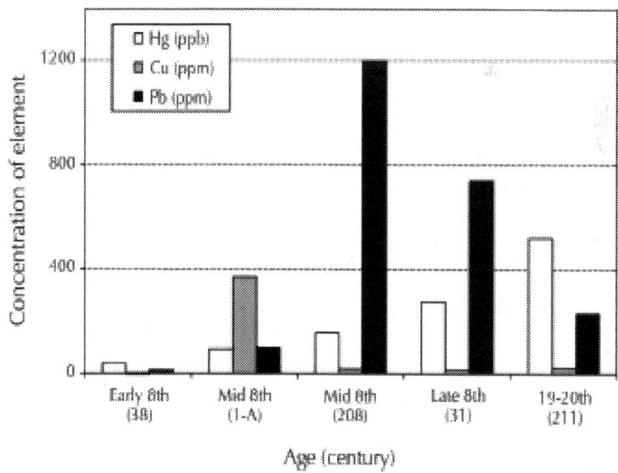

Figure 5: Hg, Cu, and Pb contents in samples 211, 208, 38, 31, and 1-A.

Cu levels varied from 2.3 to 370 ppm, with a mean of 37 ppm. Samples from the early eighth century showed that the background level of Cu was around 5 ppm. The highest values (>50 ppm) were observed in the samples from Site A, which was contaminated by copper slag that splashed out of the blast furnaces when the Nara Daibutsu was cast.

Pb content varied markedly, ranging between 13 and 1,200 ppm. The highest Pb levels (>1,000 ppm) were observed in samples 208 and 204, which came from an ironsmith dump site dating from the early-middle eighth century. Samples 34 and 31 from a ditch running along the central avenue also had high Pb contents (700 to 900 ppm). Interestingly, the mean level of Pb in Ancient Nara (330 ppm) was considerably higher than the geochemical background level of 30 ppm around the modern city of Nara.

The levels of Fe, Mn, Co, Ni, and Zn in the soil were all rather low and fell within the ranges observed in modern natural soils in Japan (Imai et al. 2004). Specifically, Fe, Mn, Co, Ni, and Zn ranged between 5,800 to 61,000 ppm (mean 26,300 ppm), 92 to 2,300 ppm (mean 425 ppm), 4.8 to 32 ppm (mean 14 ppm), <29 ppm, and 20 to 140 ppm (mean 50 ppm), respectively. The contents of the latter four metals were all correlated with the content of Fe, suggesting that these five elements are present in the same clay minerals and other soil materials.

Pb isotope ratios were relatively similar among all the soil samples studied: $^{207}Pb/^{206}Pb$ ranged from 0.850 to 0.851 and $^{208}Pb/^{206}Pb$ ranged from 2.093 to 2.094. The Pb isotope compositions of the samples from the Naganobori mine ($^{207}Pb/^{206}Pb = 0.846$ to 0.847 and $^{208}Pb/^{206}Pb = 2.091$ to 2.092) were consistent with a previous study ($^{207}Pb/^{206}Pb = 0.848$ and $^{208}Pb/^{206}Pb = 2.091$ to 2.092; Saito et al. 2002).

Metal Pollution Resulting from Construction of the Nara Daibutsu

The level of Hg pollution can be estimated if the size of the area over which the pollution occurred is known. Heijoyo in the Nara Basin measures about 5 km by 5 to 7 km and the soil layers in the basin are typically several centimeters to 20 cm thick (Figure 4). Consequently, if 820 kg of Hg was uniformly distributed in 5 or 10 cm thick soil layers (density 1.0 g cm^{-3}) in a circular area with a 3 km radius in the Nara

Basin, then the Hg concentration would have been approximately 600 or 300 ppb, respectively. The results of this study suggest that the Hg content in ancient soil in the middle and late eighth century (around 100 ppb) was not that high (Table 1). Even samples from the twelfth to eighteenth centuries that were collected in the district adjacent to the Nara Daibutsu had Hg levels of 200 to 300 ppb. Similar Hg contents were observed in the eighth century soil samples 100 and 106 collected in the palace grounds of Ancient Nara. The other samples had relatively low Hg contents. Assuming that the background level in this area was approximately 50 ppb before construction of the Nara Daibutsu, as estimated above, it appears that although Hg levels increased considerably when the statue was constructed, Ancient Nara was not severely polluted.

There are two possible reasons that might explain the discrepancy between the estimated and measured values: (1) Hg, which is volatile, may have evaporated from the soil (Pirrone et al.2001), (2) Hg might have been transported to other areas over time. Regarding the second possibility, the district to the north of Nara Daibutsu is currently flat but it was a hollow depression between the eighth to the sixteenth centuries when it would have received debris and soils from fires that burned the wooden structures housing the Nara Daibutsu in 1180 A.D. and 1567 A.D. Nonetheless, samples collected from this area, i.e., samples 1 (middle eighth century), 2 (twelfth century), and 3 (sixteenth century), had relatively low Hg contents, implying that Hg transfer was not a main factor.

The soil in Ancient Nara had low Hg levels during the early eighth century; however, these levels more than doubled after the middle eighth century, possibly due to pollution from the construction of the Nara Daibutsu. In addition, there may have been other pollution sources, as a single point source of pollution would not explain the observed increase in background Hg levels after the mid-Nara era. It is well known that cinnabar has been used for a spell and/or preservative agent for more than 2,000 years (Naruse 1991; Mitochou Compilation Committee 2004), but it was relatively expensive and would likely have been used to a limited extent.

Regarding the evaporation of Hg, the vapor pressure of Hg is 0.23 Pa at 25°C. If the Hg vapor behaves as an ideal gas, then we can calculate the concentration of Hg (C, in moles per mole of atmosphere) in the atmosphere using the following equation:

$$0.23Pa / (101 \times 10^3 Pa) = C / (1 + C)$$

where C is 2.3×10^{-6} mole, which is more than 100 times the upper limit of the modern environmental standard in Japan, 0.02 ppm. As the molar mass of Hg is 220.59 g mol^{-1}, 4.55×10^{-4} g of Hg exists in 1 mole of atmosphere, suggesting that the inside of the Great Buddha Hall was severely polluted when gilding was undertaken and that, afterward, an appreciable amount of Hg continuously and, rather rapidly, evaporated from the soil into the atmosphere. Indeed, this evaporation would explain why less than half of the total amount of Hg used in the construction of the Nara Daibutsu was present in the soil, even as soon as the middle and late Nara period. These lines of evidence suggest that a limited area around the Great Buddha may have been severely polluted during the gilding, but that Ancient Nara was only moderately polluted with Hg since the eighth century, probably due to the construction of the Nara Daibutsu as well as the use of cinnabar in the region. The evidence does not support the hypothesis that severe Hg pollution forced the capital to be relocated from Nara to Kyoto.

High Cu contents were observed in samples 1, 2, and 3, which contained copper slag from the construction of the Nara Daibutsu. The highest level observed (355 ppm) is well above the upper limit of existing Japanese regulations (125 ppm). Conversely, samples from the other sites had relatively little Cu, and background levels were probably a few parts per million. The higher levels observed in some of the samples (i.e. >10 ppm) may have been due to the widespread use of Cu in mirrors, coins, bronze statues, and arms at the time. A more likely explanation for the presence of the grassland (instead of a forest) at Wakakusa Hill was not because of pollution, but rather because the field was burned every year as part of a New Year celebration.

In summary, our results suggest that Hg and Cu pollution accompanying the construction of the Nara Daibutsu only had a limited influence on the environment.

Pb Pollution in Ancient Nara

Analysis of soils from Ancient Nara suggests that traces of ancient human activity are indeed evident in the environment. Both the extent of the contamination and the nature of the contaminants are generally consistent with the features of human activity in Ancient Nara. The contamination in samples 31 and 34 from the ancient ditch is considered to be due to urban pollution. According to historical documents, the ditch was quite dirty and contained the remains of humans, cattle, and horses, suggesting that people may have indiscriminately disposed of waste in these ditches (Archaeological Institute of Kashihara 2012a, [b]). Further, although water from the Saho River was diverted into the ditch, it was too weak to flush out the sewage efficiently (Figure 1). Hg, Pb, Fe, and Cu were also materials that were in common use in the Nara period. Hg and Pb were used for pigments, enamel, and/or lead glass; Cu and Sn were used for producing bronze coins and statues. Malachite is a copper carbonate hydroxide mineral which was used as a green pigment and Fe oxide (hematite) was used as a red pigment (Kitano 2013).

Comparing the Hg, Cu, and Pb contents of Ancient Nara to modern standards reveals that only Pb (over 330 ppm) exceeded modern levels (15 ppm for Hg, 125 ppm for Cu, and 150 ppm for Pb). The particularly high contents observed in samples 208 and 204 could be related to the high Pb content at site C. Indeed, the observed level of Pb contamination is considered harmful for human health. For example, if you inhaled 150 mg of soil per day that was contaminated with 200 ppm of Pb, and if the absorption efficiency was 40%, then 100 mg of Pb would accumulate in your body over a 20-year period and potentially cause lead poisoning (Yamada 1977). Thus, although it has been postulated that people living near blacksmiths in Ancient Nara may have suffered from lead poisoning, these smiths employed mainly iron, which means that it was unlikely that they were responsible for the lead contamination in the soil. However, further investigation is needed to determine whether Pb contamination was observed throughout the city.

The various isotopes of Pb provide an ideal tool for characterizing the original source of heavy metal pollution because the isotopic ratios are not measurably influenced by physical or chemical fractionation processes. Thus, when the $^{207}Pb/^{206}Pb$ and $^{208}Pb/^{206}Pb$ ratios of the Ancient Nara soil samples were compared to the ratios observed in various foreign artifacts (mirrors and swords from China and Korea) or in samples from various Japanese mines, the results show that the most plausible origin of the Pb in Ancient Nara was the Naganobori mine. Figure 6 shows that when the Pb isotope ratios of the soil samples were plotted against those of the Naganobori mine, the curves were similar. This mine was thus undoubtedly the source of Cu for the production of the Nara Daibutsu. Indeed, this dependence on the mine is documented in historical records, namely, on narrow strips of wood upon which official messages were written during the Nara Period (Mitochou Compilation Committee 2004): this is also evidenced by the higher relative abundance of Ag and As (Hatanaka 2003). This study reconfirmed that the Pb isotope ratios in sample 1-A, which contained drops of copper from the construction of the Nara Daibutsu, were similar to those in samples obtained from the Naganobori mine. In addition, the Pb isotopic value is the same as that in all the other solid samples in Ancient Nara. We therefore conclude that all of the Pb contamination in Ancient Nara originated from the Naganobori mine. The extraction of ore at Naganobori contributed considerably to the production of the Nara Daibutsu. However, the mining activity associated with the construction of the Nara Daibutsu may also have generated Pb pollution in the capital city, even in ancient times.

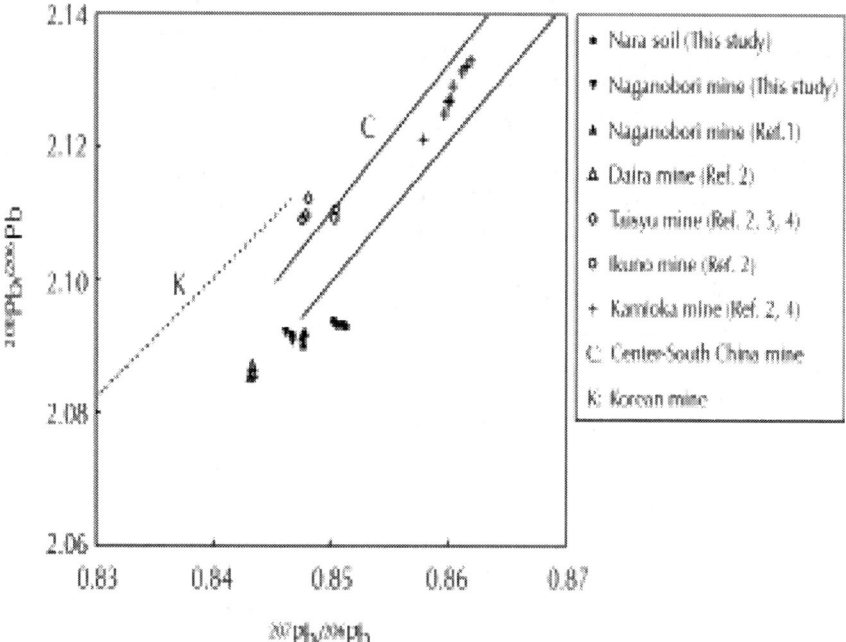

Figure 6: Pb isotopic ratios (208Pb/206Pb versus 207Pb/206Pb) in Ancient Nara soil samples and reference ore deposits. C (area between solid lines) represents bronze mirrors brought from China during the Eastern Han Dy-

nasty and the Three Kingdoms era (25 to 280 A.D.) with Pb originating from central to southern China; K (dashed line) represents a bronze mirror and swords brought from the Korean Peninsula during the Yayoi era (10 B.C. to 3 A.D.) with Pb originating from the Korean Peninsula (Saito et al. 2002). Data from (1) Saito et al. 2002; (2) Sasaki et al. 1982; (3) Mabuchi and Hirao 1982; and (4) Mabuchi and Hirao 1987.

Although it is currently difficult to identify the exact source of Pb contamination, several possible sources exist. A number of studies on Pb isotopes have examined bronze coins (Mabuchi et al.1982; Saito 2001a; Saito et al.2002). However, it is considered unlikely that these coins would have increased the Pb content of environmental samples, such as the soils in urban areas because of distribution amount. Based on an analysis of the inorganic pigments used to decorate treasures at Shosoin, the Imperial Repository constructed to prevent damage arising from the humid Japanese climate, lead-based pigments may have been relatively popular (Naruse 2004). Although it was reported that lead carbonate (lead white) was not produced in Japan and that it was imported from China during the Nara era (Winter 1981), lead chloride produced in Japan was widely used for white paint (Naruse 1992; Winter and Emile 1988). According to a production report from a government-run workshop of Buddhist sculptors during the Nara Period in 734 A.D., red lead was made from metallic lead in order to produce lead glass (Naruse 1991). Lead isotope analysis of tricolored glaze dating from the Nara Period showed that the Pb isotopic composition was comparable to that from the Naganobori mine (Saito 2001b). Further study is required to understand the extent of environmental pollution in this ancient civilization.

Implications of Pollution in the Ancient City

During the seventh and eighth centuries, many religious structures, such as temples and statues, as well as tumuli, were constructed in Japan. The material requirements of large bronze statues were considerable, as exemplified by the Nara Daibutsu; and the tumuli of the previous era were associated with large numbers of terra-cotta figures, whose firing required large quantities of wood fuel. Relocations of the capital city were also frequent and entailed extensive utilization of wood resources for the construction of new buildings.

The end of the burial tradition and the beginning of the Buddhist practice of cremation occurred in the eighth century. The first person to be officially cremated in Japan was the Buddhist priest Dosho in 700 A.D., and the first emperor to be cremated was Jito in 702 A.D. According to historical records, the government recommended restraint regarding the construction of large tumuli in the seventh century (personal communication, Dr. Kinoshita). After the relocation of the capital to Ancient Kyoto in 794 A.D., no more large bronze statues were built around Nara and Kyoto. Japanese society had already started to shift away from mass consumption to a more sustainable system during the Nara period.

CONCLUSIONS

We investigated metal pollution in the eighth century city of Ancient Nara, the first large and international city in Japan. We found that the pollution arose due to urban activity and the construction of the Nara Daibutsu and concluded the following:

- Urban activity increased the levels of Hg, Cu, and Pb in soils.
- Hg and Cu pollution accompanied the construction of the Nara Daibutsu, but the influence of this pollution was limited to a section of the city.
- At several sites, the soil was polluted with enough Pb to harm human health, with some of these values exceeding modern Japanese safety standards. The isotopic composition of Pb in these soils indicated that it originated mainly from the Naganobori mine. The mine made a major contribution to the founding of the Nara Daibutsu and was a source of some of the observed lead pollution.

AUTHORS' CONTRIBUTIONS

HK proposed the topic and conceived and designed the study. SY, KY, and HK collected samples for the study. SY, KY, TO, GS, and MI contributed to the chemical and isotope analysis. All authors read and approved the final manuscript.

ACKNOWLEDGEMENTS

We sincerely thank Dr. A. Watanabe and Dr. M. Jinno at the Nara National Research Institute for Cultural Properties, Dr. F. Sugaya, Dr. K. Saito, Dr. W. Kinoshita, and Dr. T. Fukunishi at Archaeological Institute of Kashihara, Nara Prefecture, Mr. M. Kanekata and Mr. T. Yasui at the Buried Cultural Property Investigation Center of Nara, and Mr. Y. Ikeda at the Culture Exchange Hall of Naganobori Mining for providing archeological soil samples and related data for this study. We also acknowledge Prof. K. Takemura and Dr. Y. Shitaoka of the Institute for Geothermal Sciences, Kyoto University for the instruction of archeological studies. Also, we would like to express our sincere appreciation to two anonymous reviewers for improvements to the manuscript. This study was supported by a grant-in-aid for scientific research, 22224009 (H. Kawahata), from the Japan Society for the Promotion of Science.

This paper presents quantitative data on Pb isotopes and toxic metals in samples of ancient soil and remains collected from archeological sites in Ancient Nara in order to evaluate the effect of urban activity on ancient metal pollution. Although it has been proposed that pollution resulting from urban activities (especially construction of the Nara Daibutsu) was responsible for the end of Ancient Nara and relocation of the capital, no scientific verification has been conducted to date. The Great Buddha statue is the largest bronze statue in the world and is a registered UNESCO World Heritage site, and it is well known among Japanese and foreign tourists. Therefore, the heavy metal pollution would hopefully receive much attention. This research has important implications for building a sustainable society in the future.

REFERENCES

1. Archaeological Institute of Kashihara (2012a) Hachijo North ruins. In: Archaeological Report, Archaeological Institute of Kashihara. Nara: Nara Prefecture. pp 73-94

2. Archaeological Institute of Kashihara (2012b) Hachijo North ruins. In: Reference document for information session at Hachijo North ruins. Nara: Archaeological Institute of Kashihara, Nara

Prefecture. pp 1-4

3.   Board of Education of Nara City (2011) Archinological survey of the district to the north of the Nara Daibutsu – sites of historical importance near Todaiji Temple. 1–3.Board of Education of Nara City Report in December, Nara City (in Japanese)

4.   Hall JW (1993) The Cambridge history of Japan. Ancient Japan, vol. 2. Cambridge University Press, Cambridge. p 650

5.   Hatanaka A (2003) Discussion of copper produced at Naganobori written on mokkan. Mokkann Gakkai 25:1-30 (in Japanese)

6.   Imai N, Terashima S, Ohta A, Mikoshiba M, Okai T, Tachibana Y, Togashi S, Matsuhisa Y, Kanai Y, Kamioka H, Taniguchi M (2004) Geochemical map of Japan. Geological Survey of Japan. https://gbank.gsj.jp/geochemmap/index_e.htm. Accessed 6 Jun 2014

7.   Ishino T (2004) Production of Nara Daibutsu. Komine Publishing Company, Tokyo. (in Japanese)

8.   Japan Broadcasting Corporation (2012) Nara and Asuka eras. In: Japanese history. Tokyo: NHK. p 128

9.   Japan Oil, Gas and Metals National Corporation (2010) Rare metal handbook. Kinzoku-Jihyo, Tokyo.

10.   Kawahata H, Nohara M, Aoki K, Minoshima K, Gupta LP (2006) Biogenic and abiogenic sedimentation in the northern East China Sea in response to sea-level change during the Late Pleistocene. Glob Planet Change 53:108-121

11.   Kitano N (2013) Traditional view on Bengara pigment (Hematite, α-Fe[2]O[3]) coating on wooden architectures in Japan. Yuzankaku, Tokyo.

12.   Kito H (2000) Jinkou kara yomu Nihon no rekishi (History of Japan based on the Population). Kodansya Ltd., Tokyo. (in Japanese)

13.   Konishi M (2002) Estimation of Au and Hg on the construction of the Nara Daibutsu in the Todaiji Temple. J Hist Metrology 24:91-110 (in Japanese)

14.   Mabuchi H, Hirao Y (1982) Lead isotope ratios in dotaku (bell-like object) excavated in Japan. J Archeological Soc Nippon 68:42-62 (in Japanese)

15.   Mabuchi H, Hirao Y (1987) Lead isotope ratios of lead ores in East Asia. J Archeological Soc Nippon 73:71-82 (in Japanese)

16. Mabuchi H, Hirao Y, Sato S, Midorikawa N, Igaki K (1982) Lead isotope ratios of ancient East Asian coins. Archaeology Nat Sci 15:23-39

17. Matsui K (2007) Collapse of Earth's system. Shincho-sensho, Tokyo.(in Japanese)

18. Committee MC (2004) Overview of history of Mitochou Town. Mitochou Town, Yamaguchi, Japan.(in Japanese)

19. Moore R (2011) Metropol Parasol, Seville by Jürgen Mayer H – review. Art and design, The Observer, Guardian. http://www.theguardian.com/artanddesign/2011/mar/27/metropol-parasol-seville-mayer-review Accessed 6 Jun 2014

20. Nara Meteorological Observatory (2014) net.go.jp/nara/kishou/fuken_kishou.htm.Accessed 6 Jun 2014

21. Nara National Research Institute for Cultural Properties (2010) Summary report of excavations of mokkans at the Heijyokyo Palace, Nara Capital City. Bull Nara National Res Inst Cult Properties 40:1-23 (in Japanese)

22. Nara National Research Institute for Cultural Properties (2012a) Excavations at the Nara Capital site and at Nara temples, Excavation in Blocks, 1 and 2, East First Ward on Third Street (no. 478, 486, and 488). Bulletin of Nara National Research Institute for Cultural Properties June:190–204 (in Japanese), Nara, Japan.

23. Nara National Research Institute for Cultural Properties (2012b) Press report on the excavation around the Suzaku-mon (main gate) (Heijo-kyo excavation No.495), 1–7 (in Japanese). Nara, Japan.

24. Naruse M (1991) Red seal and red pencil at Shosoin. Japanese Hist Yoshikawa Kōbunkan 521:84-89 (in Japanese)

25. Naruse M (1992) Lead white pigments in the Nara period. Annu Rep Office Shosoin Treasure House 14:33-58 (in Japanese)

26. Naruse K (2004) Inorganic pigments found on objects in the Shoroin. Bull Office Shosoin Treasure House 26:13-60 (in Japanese)

27. Ozaki M (2000) Geology of the Nara district. Geological Survey of Japan, Tsukuba.

28. Pirrone N, Costa P, Pacyna JM, Ferrara R (2001) Mercury emissions to the atmosphere from natural and anthropogenic sources in the Mediterranean region. Atmos Env 35:2997-3006

29. Saito T (2001) Comprehensive lead isotope analysis of Japanese coins issued in Nara to Yedo periods. Bull National Mus Japanese Hist 86:65-129

30. Saito T (2001) Lead isotope analysis of tricolored glaze in Nara period and green glaze in Heian period. Bull National MusJapanese Hist 86:199-208

31. Saito T, Takahashi T, Nishikawa Y (2002) Chemical study of ancient coins - lead isotope and metal composition of the Kocho-Junisen. IMES Discussion Paper, No.2002-J-30 (in Japanese)

32. Sasaki A, Sato K, Cumming GL (1982) Isotopic composition of ore lead from the Japanese islands. Min Geol 32:457-474

33. Shirasuga K (2002) Pollution by the Nara Daibutsu? J Soc Inorg Mater 9:537-541

34. Shotyk W, Weiss D, Appleby PG, Cheburkin AK, Frei R, Gloor M, Kramers JD, Reese S, van der Knaap WO (1998) History of atmospheric lead deposition since 12,370 ^{14}C yr BP from a peat bog, Jura Mountains, Switzerland. Science 281:1635-1640

35. Winter J (1981) 'Lead white' in Japanese paintings. Stud Conserv 26:89-101

36. Winter J, Emile J (1988) 'Lead white' in Japanese paintings II: measurements of lead isotope ratios. Kobunkazai no Kagaku 33:33-44

37. Yamada N (1977) Lead poisoning at present. Medical Library, Tokyo.(in Japanese)

Industrial Sludge Containing Pharmaceutical Residues and Explosives Alters Inherent Toxic Properties When Co-Digested With Oat and Post-Treated in Reed Beds

Lillemor K Gustavsson[1], Sebastian Heger[2], Jörgen Ejlertsson[3], Veronica Ribé[4], Henner Hollert[2], and Steffen H Keiter[2]

[1]Karlskoga Environment and Energy Company, Karlskoga 69121, Sweden

[2]Department of Ecosystem Analysis, Institute for Environmental Research, RWTH Aachen University, Aachen 52062, Germany

[3]Scandinavian Biogas Fuels AB, Holländaregatan 21A, Stockholm 11160, Sweden

[4]School of Sustainable Development of Society and Technology, Mälardalen University, Västerås 72123, Sweden

ABSTRACT

Background

Methane production as biofuels is a fast and strong growing technique for renewable energy. Substrates like waste (e.g. food, sludge from waste water treatment plants (WWTP), industrial wastes) can be used as a suitable resource for methane gas production, but in some cases, with elevated toxicity in the digestion residue. Former investigations have shown that co-digesting of contaminated waste such as sludge together with other substrates can produce a less toxic residue. In addition, wetlands and reed beds demonstrated good results in dewatering and detoxifying of sludge. The aim of the present study was to investigate if the toxicity may alter in industrial sludge co-digested with oat and post-treatment in reed beds. In this study, digestion of sludge from Bjorkborn industrial area in Karlskoga (reactor D6) and co-digestion of the same sludge mixed with oat (reactor D5) and post-treatment in reed beds were investigated in parallel. Methane production as well as changes in cytotoxicity (Microtox(R); ISO 11348–3), genotoxicity (Umu-C assay; ISO/13829) and AhR-mediated toxicity (7-ethoxyresorufin-O-deethylase (EROD) assay using RTW cells) were measured.

Results

The result showed good methane production of industrial sludge (D6) although the digested residue was more toxic than the ingoing material measured using microtox$_{30min}$ and Umu-C. Co-digestion of toxic industrial sludge and oat (D5) showed higher methane production and significantly less toxic sludge residue than reactor D6. Furthermore, dewatering and treatment in reed beds showed low and non-detectable toxicity in reed bed material and outgoing water as well as reduced nutrients.

Conclusions

Co-digestion of sludge and oat followed by dewatering and treatment of sludge residue in reed beds can be a sustainable waste management and energy production. We recommend that future studies should involve co-digestion of decontaminated waste mixed with different non-toxic material to find a substrate mixture that produce the highest biogas yield and lowest toxicity within the sludge residue.

BACKGROUND

Methane production as biofuels is a fast and strong growing technique for renewable energy. According to Eurostat [1], total production of biogas in Europe was 100 million tonnes of oil equivalents (toe), corresponding to 9% of the total biofuel production in the European Union [1]. In Sweden, total renewable energy production was 9,993 ktonnes [1], corresponding to 1,363 GWh. Biogas from landfill, waste water treatment plants (WWTP), co-digestion plants and industrial plants accounted for 22%, 44%, 25% and 8%, respectively [2]. The aim of EU council [3] states that 20% of final energy consumption should be provided by renewable sources by 2020. Although the production of biofuels is growing, this goal is probably not realistic [4].

The part of 8% industrial waste (food waste, sludge from WWTP and industrial wastes) has potential to increase since some of these type of waste possess good methane gas potential comparable with common substrates for biogas production like grass, wheat and straw [5]. However, digested WWTP sludge is highly questioned for use as fertilizer in agriculture [6-9], and industrial waste can also possess inherent toxicity inhibiting digestion process and resulting in a digested residue needing post-treatment. Co-digesting of hardly degradable and toxic material together with some other substrate has also been demonstrated to be a realistic option [10-15].

The main alternatives for a digested residue with toxic or environmental hazardous properties are combustion, composting and/ or use in less sensitive land applications such as covering of landfills. One alternative cost-efficient and low-technology demanding method, efficient in reducing nutrient content, carbons and toxicity, is the dewatering and treatment of sludge in constructed wetlands (CW) and reed beds [16-18]. More than 7,000 CW is operating in Europe and North America with increasing number in South America, Australia, New Zeeland as well as Africa and Asia [16-24]. The removal efficiency of nutrients and pollutants by CW and reed beds can be explained by the rhizosphere providing a large attachment area for both aerobic and anaerobic microorganisms[16,17,23,25-28] and as well as dewatering capacity by evapotranspiration and mechanical impact of shoots, roots and rhizome growth [17,22-24,29]. The capacity of planted beds in treating sludge from the same industrial area as in the present study, in comparison to filter beds without vegetation, has been demonstrated earlier [29,30]. Results showed that reed-planted beds were more efficient than unplanted at retaining toxicants, reducing the water-soluble toxicity [30], total organic carbon (TOC), biological oxygen demand (BOD) and chemical oxygen demand (COD) in the outgoing water [29,30].

Common reed has also demonstrated a high adaptive capacity to sewage sludge environment, and a doubling of shoot density compared to natural stands has been observed [20,24,29].

In this study, digestion of industrial sludge from Björkborn industrial area in Karlskoga containing nitroaromatic compounds, explosives and pharmaceutical residue and co-digestion of the same sludge mixed with oat was studied in parallel as well as post-treatment and dewatering of the digested sludge in reed beds. Earlier studies of the sludge used in this study industrial sludge from Björkborn industrial area in Karlskoga, showed good methane production potential during mesophilic conditions [31] but increased toxicity in the digested sludge [32-34]. Dewatering and treatment of this particular sludge demonstrated high dewatering capacity and reduced nutrient levels in outgoing water and sludge residue [29] as well as significantly reduced toxicity in outgoing water and bed material of reed beds measured with DR-CALUX, Umu-C assay and fish embryo toxicity test using *Danio rerio*[30].

The aim of the present study was to investigate co-digestion of industrial sludge from Björkborn industrial area in Karlskoga and oat. We wanted to investigate if co-digestion of substrate, considered as waste, together with common crop is suitable for biogas production. Moreover, we wanted to check if oat could alter the methane yield and toxic properties of industrial sludge by measuring methane production as well as change of cytotoxicity, genotoxicity and dioxin-like activity. In addition, we also wanted to investigate if co-digestion produced a less toxic and post-treatment demanding residue when dewatered through reed beds.

RESULTS AND DISCUSSION

Biogas Production

The results of the biogas measurements (Table 1) showed that the mixed reactor with sludge and oat (D5) produced methane gas at a level below the control reactor with oat (D4) but possessed a methane gas potentially higher than the sludge reactor (D6). However, gas production in reactor D6 (Table 1) was almost twice as high compared to methane production of the same sludge in a former study where a gas production of 2,000 ml/day at 37°C (60% methane) was achieved using the same organic loading rate (OLR) of 3 g VS/L reactor/day [31].

Table 1: Biogas and methane yield

Reactor	Substrate	OLR (g VS/L/ day)	CH4 (%)	Gas production (ml/day at 37°C)	Specific gas production (ml methane/g VS at 0°C)
D4	Oat	6	51	17,500	300
D5	Oat+sludge	3 + 3	53	14,000	270
D6	sludge	3	63	4,000	180

Gustavsson et al.

Gustavsson et al. Environmental Sciences Europe 2014 26:8, doi:10.1186/2190-4715-26-8

The measured toluene, benzene, ethylbenzene and xylen concentrations during routine controls (ALS; Table 2) exceeded the limits for land application and land use in Sweden [35]. These compounds are mainly degradation products of 2,4,6-trinitrotoluene (TNT), di-nitrotoluens, nitrobenzoic acids and a range of other compounds used for the manufacturing of explosives, pharmaceutical and chemical intermediates [32-34].

Table 2: Content of organics and metals in a month sample of undigested sludge

Organic compounds		Metals	
Parameter	(mg/kg TS)	Parameter	(mg/kg TS)
TS (%)	20	As	<3
VS (% of TS)	68.4	Ba	79.8
AOX	84	Be	0.264
Benzene	0.76	Ca	25,600
Toluene	260	Cd	<0.1
Etylbenzene	0.27	Co	4.17
Xylene	13	Cr	74.3
Di-etylftalat	4.1	Cu	25.2
Di-n-butylftalat	0.19	Fe	63,000
Di-n-pentylftalat	32	Hg	<1
Di-(2etylhexyl)ftalat	3.1	Li	0.517
PAH (sum)	<4.3	Mn	85
4-Nonylphenol	<0.25	Mo	4.9
RDX	0.43	Na	2,050
HMX	0.45	Ni	4.08
TNT	0.82	P	24,600
		Pb	29.2
		S	17,300
		Sr	34.3
		V	21
		Zn	99.8

From Björkborn industrial area collected during routine sampling.

Gustavsson et al.

Gustavsson et al. Environmental Sciences Europe 2014 26:8, doi:10.1186/2190-4715-26-8

However, the OLR of 3 gVS/L used in the present study did not inhibited the digestion process and resulted in gas production comparable with ordinary municipal sewage sludge in Sweden with an average gas production of 160 to 350 m^3 CH_4/tonne VS [36,37]. OLR in D6 is half of OLR in D5, resulting in twice as high hydraulic retention time (HRT) which may give the microorganisms time for adaptation but can also explain the lower methane yield. The lower HRT in D5 (Table 2) may also explain lower toxicity. A previous study [38] found that decreasing HRTs results in higher feeding and outgoing flow rates and, consequently, rapid withdrawal of toxic intermediates and less accumulation of inhibiting intermediates [38]. Intermediates can originate from the degradation of aromatic amino acids [39].

A decreased HRT can prevent the accumulation of toxic substances and inhibition of the digesting process [10, 38], but as a consequence, the methane yield would be reduced [38]. Instead, co-digestion could be a promising alternative option. Olive mill waste (OMW) possesses a high energy potential (45 to 220 g of COD/L) but also a low pH, alkalinity and nitrogen content; additionally, a lipophilic fraction and phenolic compounds are present. These characteristics make this substrate toxic and complex to degrade during anaerobic conditions [14]. However, it is increasing the methane yield when co-digested with manure (Table 3). Earlier studies (Table 3) of co-digesting different substrates revealed promising biogas production and showed high methane yield.

Table 3: Examples of previously performed studies of co-digestion

Substrate	(Mixed ratios v/ v)	HRT (days)	OLR (g VS/L/ day)	Methane yield (ml CH4/g VS/day)	References
OMW + cattle manure	75% + 25%	13	3,4	700 to 1,000	Angelidaki and Ahring [10]
OMW + pig manure	69% + 31%	6	2.9 + 2.6	2,700	Sampaio et a.l [14]
Sewage sludge + potato waste	44% + 56%	20	2.7	600	Murto et al. [13]
Industrial waste + pig manure	17% + 83%	30	2.6	800	Murto et al. [13]

Industrial waste + pig manure + slaughterhouse waste	17% + 71% + 12%	28	3.1	900	Murto et al. [13]
Industrial waste + pig manure + slaughterhouse waste	17% + 66% + 12% + 5%	36	2.6	1,000	Murto et al. [13]
Sewage sludge + marine dredgings + municipal refuse	20 + 5% + 75%	36 (batch)		900 to 1,200	Chan et al. [11]
OMW + piggery effluent	83% + 17%	6 to 7	3.5	1,300	Marques [12]
Oat + sludge	50% + 50%	28	6	500	Present study

Gustavsson et al.

Gustavsson et al. Environmental Sciences Europe 2014 26:8, doi:10.1186/2190-4715-26-8

Co-digesting of starch-rich and ammonia strong wastes obtained gas yields comparable with yields obtained in the present study when co-digesting sewage sludge and potato processing industrial waste [13]. Co-digesting of manure, slaughterhouse and agricultural waste revealed higher gas yield (Table 3) with higher diversity of substrate [13]. This is consistent with the study performed by Chan et al. [11] who tested co-digestion of sewage sludge and marine dredgings mixed with municipal refuse at 13 different ratios [11]. Additionally, results in this study are strengthened by other studies which also found that co-digesting enhanced biogas production compared to digestion of single material such as high-strength COD substrate [10,13,14].

The methane yield in this study is lower than in many other co-digestion studies (Table 3). The explanation can be the presence of hardly degradable and toxic nitro-aromatic compounds (Table 2). The present study of 4,000 ml biogas/day and 300 ml CH_4/g VS added (Table 1) are confirmed by earlier studies demonstrating digestion of nitroaromatic compounds where methane gas production of 2,300 ml/day was achieved using a nitro-benzene loading rate of 30 mg/L/day [40]. In another study, methane yields between 116 and 209 ml CH_4/gVS L^{-1} were obtained by adding p-nitrophenol [38].

A low methane yield can also be explained by low carbon/nitrogen (C/N) ratio. The C/N ratio of 1.7 (Table 4) makes the industrial sludge unfavourable for digestion since a ratio of 16 is required to balance the anaerobic degradation between accumulation of volatile fatty acid (VFA) during digestion with high C/N ratio or accumulation of ammonia with low C/N ratio [10,13]. C/N ratio in oat mixed with sludge was higher (3.6) but still unfavourable according to [13]. On the other hand, if the activity of methanogenic bacteria is low, less of the proteins will be degraded to free ammonia ions, inhibiting the digestion process [5].

Table 4: Nutrient and organic change in the sludge and sludge + oat reactor

(mg/L)	Ntot	TOC	Ptot	NH4	C/N
Ingoing sludge D5	103	369	4.2	46.7	3.6
Digested D5	110	215	1.5	89.0	2.0
Sludge residue D5	28	275	1.4	0.4	9.9
Bed material D5	49	8	1.3	17.2	0.2
Outgoing water D5	Nm	118	1.5	2.7	0.2
Red (%)		68	65.0	94.2	-0.2
Ingoing sludge D6	811	1,340	10.7	800	1.7
Digested D6	777	1,290	13.4	746	1.7
Sludge residue D6	439	350	3,4	0,31	0,8
Bed material D6	123	373	32.3	24.3	3
Outgoing water D6	347	151	10.1	164	0.4
Red (%)	57	89	5.6	79.5	1.6

Before and after digestion and dewatering through reed beds after 3 months. Reduction (%) of nutrients and organics is calculated between ingoing sludge and outgoing water from reed beds. Italic number means increase. nm, not measured.

Gustavsson et al.

Gustavsson et al. Environmental Sciences Europe 2014 26:8, doi:10.1186/2190-4715-26-8

Post-Treatment in Reed Beds

Reduction of nutrients and carbon throughout the three months of digestion and dewatering in reed beds is shown in Table 4. The results show a high reduction of ammonia for both reed bed lines (Table 4) despite the short treatment time. This result is consistent with the findings of an earlier study of dewatering the sludge from Björkborn industrial area [29]. The authors found that reed beds were able to reduce COD and TOC to more than 90%, BOD and total nitrogen (N_{tot}) to more than 80% and total phosphorous (P_{tot}) to over 85% during the growth period (April to October). During the resting period (November to March), reduction of COD, BOD, N_{tot} and P_{tot} decreased to 66%, 28%, 35% and 55%, respectively [29].

Additionally, several other studies have shown a high post-treatment capacity of reed beds. BOD removal efficiency of 63% to 79% independent from the season or age of the system was reported[41]. Moreover, other studies demonstrated removal efficiency of nutrients, nitrogen, phosphorus, BOD and total suspended solids (TSS) using constructed wetlands [21,42].

Toxicity Tests

Microtox

The result of the present study showed an increased toxicity and an accumulation potential of non-reduced nitroaromatic compounds of the industrial sludge in reactor D6 during digestion. This is consistent with earlier studies investigating the same sludge where an increased acute toxicity and decreased cell vitality were measured after exposure to extracts of digested sludge using the microtox [31] and neutral red assay [32], respectively. Reed beds containing industrial sludge (D6) showed a decreased toxicity in sludge residual and bed material, suggesting that some of the compounds causing toxicity were transformed to water-soluble compounds and rapidly transported through the reed beds, ending up in the outgoing water. A large portion of those compounds, which are not trapped in the bed material or flushed out with the outgoing water, were probably degraded in the reed beds as shown in earlier studies of dewatering sludge from Björborn industrial area [30].

A contrary result was observed by testing a mixture of sludge and oat from D5. Ingoing sludge of D5 was less toxic compared to ingoing sludge of D6 (Figure 1). The differences between the reactors (Figure 1) are larger than the dilution effect of oat by a factor of 2 (Table 1). Additionally, sludge from D5 showed decreased toxicity after digestion in opposite to digested D6 sludge demonstrating increased toxicity.

The comparison of both reactors after digestion demonstrates a larger difference in toxicity that cannot be explained by dilution effect alone. Acute toxicity could not be detected in outgoing water from reed beds of D5. Sludge residue and bed material trapped toxic compounds, demonstrated by slightly increased TU values in D5 (Figure 1, Table 5).

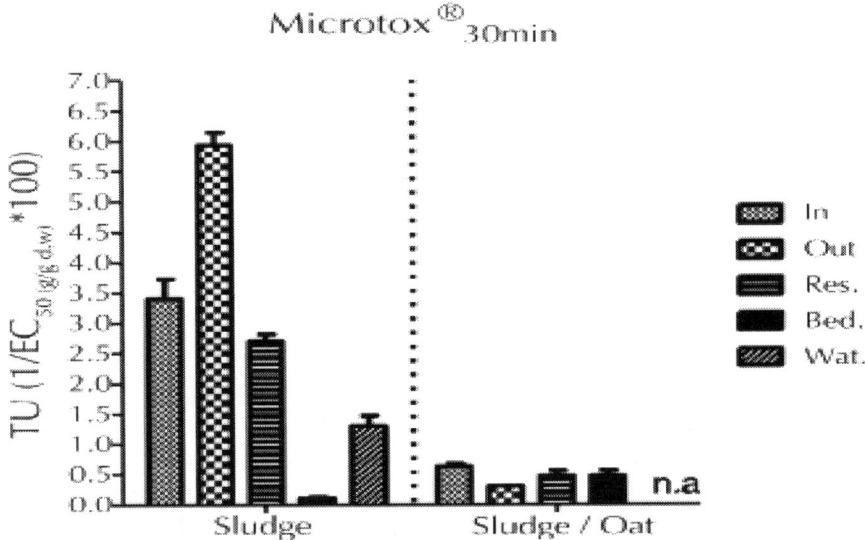

Figure 1: Toxic units (TU) based on EC50 (mg/g dw). Data are given as mean and 95% confidence interval. Each sample was tested in two independent replicates.

Table 5: Summary table of the result of ecotoxicity test

Sample	EROD		Umu-C induction > 1.5	Microtox TU
	EC5	EC10	LID (µl/ml)	(mg/g dw)
D5 undigested	n.a		165	0.6
D5 digested	n.a		165	0.3
D5 sludge residue	36.8	100.8	165	0.5
D5 bed material	n.a		165	0.11
D5 outgoing water	n.a		n.m	n.a
D6 undigested	n.a		82.5	3.3

D6 digested	n.a		41.25	5.9
D6 Sludge residue	20.4	37.8	n.m	2.7
D6 Bed material	68.5		n.m	0.04
D6 Outgoing water	n.a		82.5	1.3

n.a, not available; n.m, not measured.

Gustavsson et al.

Gustavsson et al. Environmental Sciences Europe 2014 26:8, doi:10.1186/2190-4715-26-8

Umu-C

In the Umu-C assay (ISO 13829), a genotoxic effect is significant if the induction factor is above 1.5 compared to the negative control. Figure 2 shows the used concentrations of the different samples reaching an induction factor of 1.5. Genotoxicity above 1.5 was detected in the undigested and digested D6 sludge with LID values of 82.5 and 41.25 µl/ml, respectively (Figure 2, Table 5). This result clearly demonstrates that the toxicity increased after digestion with only half the concentration needed to be genotoxic compared to undigested sludge. Outgoing water from reed beds treating D6 sludge showed higher LID values indicating that genotoxic compounds may have been adsorbed in bed material or degraded.

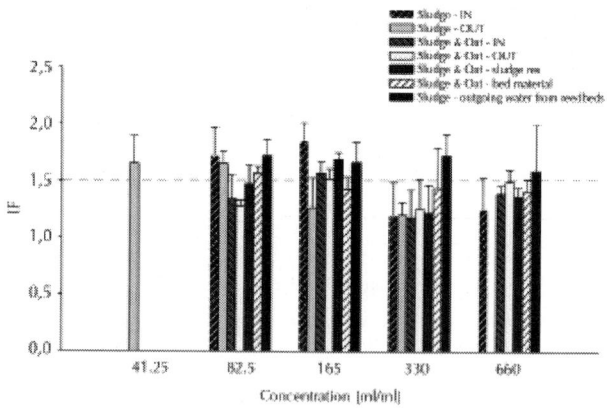

Figure 2: Mean and std dev of concentrations that caused induction factor above 1.5 compared to control. $n = 3$.

The mixed material with sludge and oat from reactor D5 showed unchanged genotoxic activity before and after digestion but decreased LID values in bed material which points at adsorption of genotoxic compounds in the bed material (Figure 2).

The genotoxic properties of this industrial sludge have been demonstrated before. A previous study [32] showed significant genotoxic potential in the digested sludge compared to undigested, tested in the comet assay with RTL-W1-cells. A former study of large scale anaerobic treatment of sludge from Björkborn industrial area demonstrated an increased genotoxic activity during treatment from induction factor 1.5 to induction factor 2.8 in the Umu-C assay [34]. This is higher as the genotoxic potential of D6 in the present study, although the pattern is the same, an increasing induction factor during anaerobic treatment (Figure 2, Table 5).

Different publications describe the genotoxic properties of nitro-aromatic compounds such as TNT, nitrobenzoic acids, nitrobenzenes and degradation products [43-46] and increasing toxicity with increasing number of nitro-groups [34,47]. Additionally, literature has also shown higher toxicity with nitro-substituted aromatics compared with their corresponding amines [38,48-50]. This may explain the increased genotoxicity, although weak, by the presence of unreduced nitro-aromatics within the sludge used in this study. 7-Ethoxyresorufin-O-deethylase (EROD) EROD activity could only be detected in three samples (Figure 3, Table 6). The EC_5 was comparably low in D5 Res (36.8 mg/ml), D6 Res (20.4 mg/ml) and in D6 Bed (68.5 mg/ml). All other samples did not induced EROD activity.

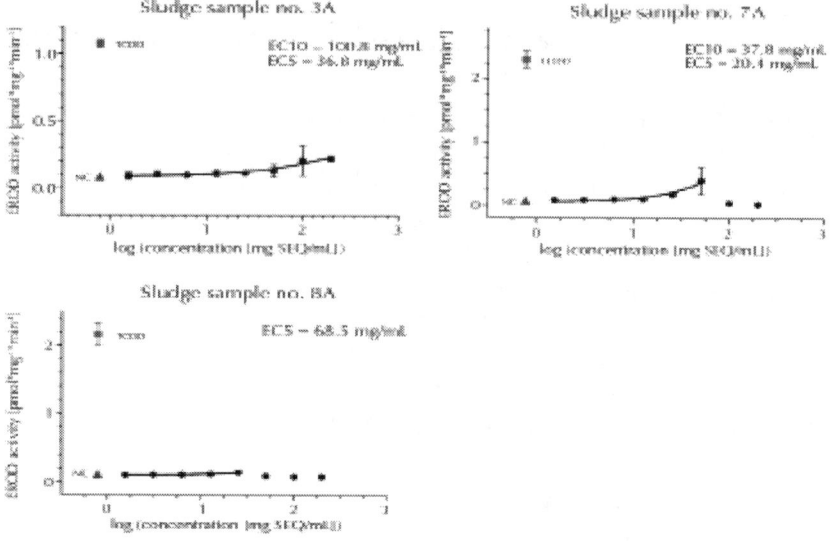

Figure 3: EROD activity detected in three samples. EC5 and EC10 in sludge residue of reactor D5 (no. 3a), in sludge residue of reactor D6 (no. 7a) and bed material in reed beds treating sludge from reactor D6 (no. 8a).

Table 6: Samples, extraction solvents and analysis used in this study

Sampling point	Sample	Analysis			
		TOC, NH4+, Tot N, Tot P	Microtox	Umu-C assay	EROD assay
Ingoing sludge	IN	Water phase	Water phase	Water phase	Toluene (Soxhlet)
Digested sludge	Out	Water extract	Water extract	Water extract	Toluene (Soxhlet)
Sludge residual	Res	Water phase	Water phase	Water phase	Toluene (Soxhlet)
Bed material	Bed	Water extract	Water extract	Water extract	Toluene (Soxhlet)
Outgoing water	Wat	Native water	Native water	Native water	Toluene

The material used in the analysis is in bold in the first column. The first row in bold and italics describes the analysis used in the study, and the cells from the second row and column describes the extracts used.

Gustavsson et al.

Gustavsson et al. Environmental Sciences Europe 2014 26:8, doi:10.1186/2190-4715-26-8

After 3 months of loading, detectable levels of EROD inducers could only be found in sludge residue on top of the reed beds from both D5 and D6 and in the bed material from D6. However, the levels are very low, and it was impossible to calculate Bio-TEQ values. Mesophilic digestion can increase EROD activity compared to undigested material. In a former study by [33], the same sludge obtained from the same manufacturing area as in this study, demonstrated three to six times higher levels of EROD activity in the digested sludge than in the incoming flux [33]. Additionally, earlier studies of methanogenic digestion of household waste showed that acid anaerobic conditions [51,52].

The identity of these EROD inducers was neither clarified in the former study by [33] nor in this present study. Additionally, it has been shown that a variety of different compounds apart from the well-known dioxins and polychlorinated biphenyls (PCBs) can induce EROD activity. For instance, conversion of proteins like tryptophan to indole acetic acid (IAA) and transformed compounds like indole-3-carbinole (I3C) and indolo-3.2- -carbazole (ICZ) demonstrated 1×10^2 and 1×10^5 times higher AhR binding affinity than the parent compound, respectively [53,54]. Additionally, several phytochemicals including caffeic acid, chlorogenic acid, diosmin, ferulic acid and resveratrol showed both inhibition and induction of EROD [55].

CONCLUSIONS

Digestion of sludge from Björkborn industrial area (D6) resulted in a methane production of 180 ml CH_4/g VS, comparable with methane production of WWTP sludge. However, the digested residue was more toxic than the ingoing material measured using $microtox_{30min}$ and Umu-C.

Co-digestion of toxic industrial sludge and oat (D5) showed higher methane production (270 ml CH_4/g VS) despite the fact that just half of the HRT was used. Moreover, the digested residue was significantly less toxic than the sludge residue of D6. The differences in toxicity (i.e. Microtox) cannot be explained by dilution effects (OLR and HRT) as discussed in the 'Results and discussion' section and in this section. This clearly demonstrates the benefits of co-digestion of industrial sludge together with oat. Furthermore, dewatering and treatment in reed beds showed low and non-detectable toxicity in reed bed material and outgoing water as well as reduced ammonium (NH_4^+), N_{tot} and TOC. Moreover, toxicity of the dewatered D5 sludge on top of the reed beds were significantly lower than corresponding D6 for all three bioanalytical tests used in this study.

A less-contaminated waste stream demands less energy and monitoring during treatment. Therefore, digestion of sludge resulting in a less toxic residue, with a shorter and less complex post-treatment is the most cost-efficient option. We have demonstrated that co-digestion of industrial sludge with oat fulfilled that requirement. Additionally, dewatering and treatment of sludge in reed beds can be recommended as a post-treatment method of digested sludge.

Future studies should involve co-digestion of this industrial sludge or other waste mixed with different straw and grass in different proportions in order to find a substrate mixture that produces the highest biogas yield and lowest toxicity within the sludge residue. Using waste as a substrate in a sustainable way can also increase the possibilities to reach the aim of the EU council [3], stating that 20% of final energy consumption should be provided by renewable sources by the year 2020.

METHODS

Substrates and Inoculums Used in the Study

Industrial sludge used as feeding material to the bioreactors was collected from the wastewater treatment plant of Björkborn industrial area, (Karlskoga, Sweden). Approximately 25 kg of dehydrated sludge was collected and carefully mixed and aliquots were stored in 1 L polyethylene bottles at -20°C. Total solid content (TS) of the sludge mounted 15.8, and volatile solid (VS) was 65.8% of TS. The oat was received from Söderslätts Spannmålsgrupp and milled to a grain size of 1 mm prior storage at room temperature in 1 L polyethylene cans during the experimental period. TS of the mounted 96%, and the VS of TS was 97.3%. Inoculum for the laboratory digesters consisted of digested sewage sludge from Reningsverket Nykvarn (Linköping, Sweden) and cow manure from Swedish dairy farm (Hags gård, Rimforsa, Sweden).

Digestion of Sludge and Oat in Bioreactors

In the study, two digesters and one control were operated at 37°C with 20 days of HRT for the co-digestion of industrial sludge, milled oat (D5) and industrial sludge (D6), in parallel with a control reactor fed with milled oat (D4). The control reactor was operated according to the same protocol as the experiment reactors. Each digester contained an active liquid volume of 4 L and was equipped with a tube for feeding substrate/withdrawal of reactor material, a gas outlet and a central placed impeller (Ø:70 mm) for mixing. Mixing was performed in 15 min intervals four times a day and for about 10 min in connection to feeding by use of a servomotor (MAC050-A1; All motion technology, New York, NY, USA) at 500 rpm.

Digester D5 was inoculated with 2.5 L digested sewage sludge and 500 g of cow manure followed by the addition of 1 L deionized water. The following day, 200 g of digester liquid was withdrawn followed by the feeding of 200 g of a substrate blend consisting of the industrial sludge (2.0 g VS/L/day), oat 0.25 (g VS/L/day) and deionized water. The same feeding procedure was done for 28 days. To start reactor D6, digested sludge (200 g per day) from reactor D5 was collected for the last 10 days and transferred to a digester. At an active volume of 2 L in D6, 1 L digester liquid was transferred from D5 to D6.

Digester D5 was fed with the same substrate blend until an active volume of 4 L was resumed. The loading rate was allowed to increase with 0.5 kg VS/L/day every 5 days until 3 g VS/L/day was reached first with the industrial sludge and then 3 g VS/L/day with oat. Digester D5 was then fed with this substrate blend for 60 days when the experiment was terminated. Digester D6 was fed with the industrial sludge (2 g VS/L) and deionized water until an active volume of 4 L was resumed. The loading rate was then allowed to increase with 0.5 kg VS/L/day every 5 days until 3 g VS/L/day was reached. This loading rate was kept for the remaining experimental period of 60 days.

Gas production was recorded on daily basis. The methane content of the produced gas was measured once a week. The produced gas was collected in a balloon during 24 h, and the gas composition was determined using a portable gas detector (Gas data, GFM series, Whitley, Coventry, UK). Analyzed gases, besides CH_4, were CO_2, O_2 and H_2S. Samples were also taken from the reactor liquid; concentrations of individual VFAs (acetic, propionic, butyric, isobutyric, capronic, isocapronic, valeric and isovaleric acid) were determined twice a week by GC-FID [56], pH at least twice a week and TS/VS once a week following the protocols from Swedish Standard SS-EN 12176 and SS 028113, respectively.

Post-Treatment in Reed Beds

Two lines of reed beds treating sludge from D5 and D6 were constructed. Three beds with 32 cm of sand and gravel, from top to bottom 10 cm of sand, 9 cm of coarse sand, 7 cm of gravel and 6 cm of small stones (Figure 4) with vertical flow, were constructed indoors with a volume of 25 L and a upper surface area of approximately 700 cm² (Figure 1). The beds were planted with common reed (*Phragmites australis*) and kept indoors under controlled conditions with 300 mmol photons/m² s⁻¹ and loaded with sludge (diluted to 1.2% dry weight, 1.0 L/day); retention time was 2 h. The loading was continued within 3 months.

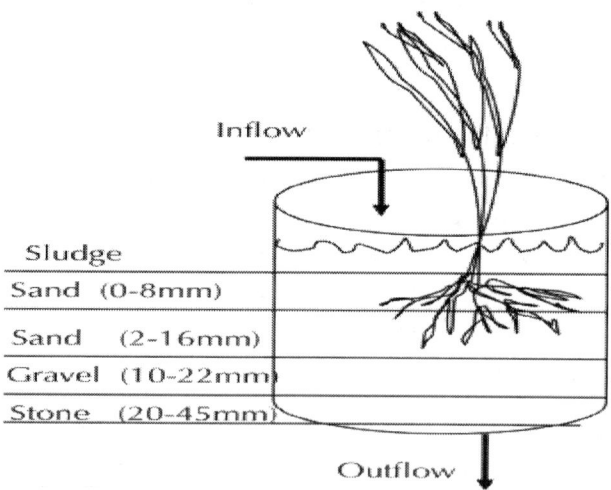

Figure 4: Schematic description of reed beds.

Sampling, Preparation, Extraction and Cleanup

All tests in this study were performed on the undigested sludge, digested sludge and outgoing water (Figure 5) of the last month sample after 90 days of digesting. Outgoing water was collected daily and pooled into a monthly sample and stored at -20°C prior to analysis. Sludge residue on top of the reed beds and bed material (Figure 5) was collected at the end of the study.

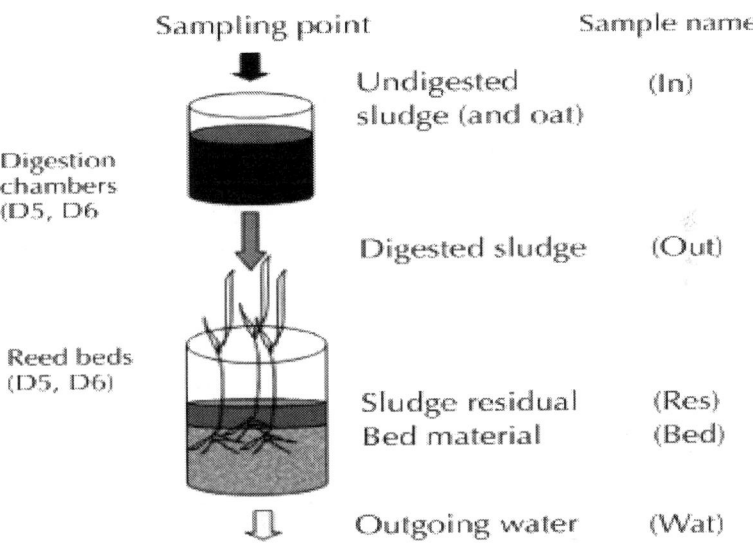

Figure 5: Experimental design and sampling points. Tests in this study performed on the undigested sludge, digested sludge and outgoing water.

Approximately 500 g of the bed material and sludge residual was collected from the upper part of the beds consisting of sand. Only the upper layer was collected for the experiments based on the assumption that most of the compounds will be trapped in the upper part of the beds due to the large specific area of this bed material compared to the coarse material below the upper part. In brief (Figure 5), ingoing sludge (In), digested sludge (Out), sludge residue (Res), bed material (Bed) and outgoing water (Wat) were extracted with water and toluene (Table 6) according to a former study in order to estimate both the bioavailable toxicity and the total toxic potential, respectively [30].

For toxicity testing of sludge and bed material, a Soxhlet extraction was conducted (24 h, 3 cycles/h) by using toluene (Riedel-de Haën, >99.8% (GC)), Envisolv according to [57]. The toluene extracts for the EROD assay were cleaned up using a multi-layer silica column in order to isolate the persistent lipophilic compounds according to the protocol shown in [58] and [59]. The silica column consisted from the bottom of 1 cm³ copper powder to precipitate the remaining sulphate and 5.3 cm³ KOH, 0.88 cm³ neutral silica, 5.3 cm³ 40% H_2SO_4, 2.65 cm³ 20% H_2SO_4, 1.76 cm³ neutral silica and 1.76 cm³ $NaSO_4$ (monohydrate). The remaining fraction contained persistent dioxin-like compounds and included, e.g. PCDDs/PCDFs and PCBs and were eluted with n-hexane. The solvents were evaporated under a nitrogen stream, and the sample was transferred to DMSO (Sigma assay (GC) minimum 99.5%, Sigma-Aldrich, St. Louis, MO, USA) for the subsequent EROD assay.

The ingoing sludge, bed material and sludge residue were also prepared for analysis of toxicity, TOC, N_{tot}, NH_4^+ and P_{tot} by shaking with 1:5 proportion of deionized water for 24 h followed by centrifugation at $5,700 \times g$. The water-phase supernatant was used for testing. The outgoing water from the beds was collected, and 100 ml from each time point was pooled to a monthly sample and stored at -20°C until analysis. The water was centrifuged and tested undiluted.

Analysis of organics and metals were performed of the ingoing sludge of reactor D6 (Table 2) at the commercial laboratory, ALS Scandinavia AB, Luleå, Sweden (ALS).

Toxicity Tests

Microtox

Toxicity to *Vibrio fischeri* of water extracted samples was assessed according to the Microtox® ISO 11348–3 test protocol (2007) by using Microtox Omni™ Software (Azur Environmentals, Newark, DE, USA). The samples and two controls consisting of deionized water were adjusted to a salinity of 2 ppt. Light inhibition in the sample compared to the control was measured after 30 min of incubation. Samples were diluted by a 1:2 series, and each dilution step was prepared in duplicates. The sample concentrations tested were 80%, 50%, 33.33%, 25%, 16.67%, 12.50%, 8.33% and 6.25%. EC_{50} values (30 min) were determined from concentration-response curves. Toxic units TU (g/g dw) was calculated using the formula: $TU = 1/(EC_{50} \times 100)$.

Umu-C

Genotoxicity of water extracts from the sludge and bed material as well as outgoing water from the beds was detected using the Umu-C test with *Salmonella typhimurium* TA1535/pSK1002 according to the standard protocol ISO 13829 (2000). The bacteria were cultured in tryptone/glucose medium in 96-well plates (Labdesign, TCT, Lake Charles, LA, USA). All concentrations were tested in triplicates. As positive and negative control, 50 µg 4-nitroquinoline-1-oxide (4-NQO)/L and pure medium was used, respectively. Induction of genotoxicity, expressed as -galactosidase activity was measured as the absorbance at 420 nm after 2 h of exposure followed by 2 h post-incubation. Growth was measured as the absorbance at 600 nm. Absorbance was measured with a microplate reader (Expert 96, MikroWin 2000, Asys/Hitech, Eugendorf, Austria). The result was calculated as an induction ratio related to growth in Equation 1.

$$\text{Induction ratio} = (1/\text{Growth}_{\text{Abs600 nm}} \times (\text{Samples}_{\text{Abs 420 nm}}/\text{Control}_{\text{Abs420 nm}})) \tag{1}$$

The test was considered valid if the growth factor at a wavelength of 600 nm of exposed bacteria versus negative control was not below 0.5 and the induction ratio measured at 405 nm of the positive control was at least twice compared to the negative control. The samples were considered genotoxic if the induction factor exceeded 1.5 (exposed bacteria versus negative control) measured at 405 nm.

EROD Assay

Induction of 7-ethoxyresorufin-O-deethylase was measured in the CYP 1A expressing permanent fish cell line RTL-W1 (rainbow trout liver, *Oncorhynchus mykiss*). Cells were obtained from Dr. Niels C. Bols (University of Waterloo, Canada) [60] and maintained at 20°C in

75 cm² plastic culture flasks (TPP, Trasadingen, Switzerland) without additional gassing in Leibovitz medium (L15) supplemented with 9% foetal bovine serum (Th. Geyer, Renningen, FRG), 1% penicillin/ streptomycin (Sigma-Aldrich). Induction of EROD was measured in confluent cell monolayers in 96-well microtiter plates (TPP) with 3 to 4×10^5 cells/ml according to a previously published method [61,62]. Before exposure to the sludge extracts, cells were seeded in 96-well microtiter plates at a density of 3 to 4×10^5 cells/ml and allowed to grow at 20°C to confluency for 72 h. Subsequently, the medium was removed and the cells were exposed for 72 h with the sludge extracts water dilutions in L15 medium, negative control (L15 medium) and positive control using 100 to 3.125 pM 2,3,7,8-tetrachlorodibenzo-p-dioxin (TCDD, Sigma-Aldrich, Deisenhofen, FRG). After exposure, all plates were shock-frozen and stored at -80°C for at least 1 h until EROD measurement.

For measurement of the EROD activity, the plates were thawed for 10 min, the protein standard solution (10 to 1.25 mg/ml bovine serum albumin (Sigma-Aldrich) in phosphate buffer (0.1 M Na_2HPO_4-solution; Malinckrodt Baker, Deventer, Netherlands) adjusted to pH 7.8 with 0.1 M NaH_2PO_4-solution (Merck, Darmstadt, FRG) was added in triplicates and 100 to 3.125 nM resorufin standard (Sigma-Aldrich) in phosphate buffer was added in duplicates. The 7-ethoxyresorufin solution (100 µl, 1.2 µM, Sigma-Aldrich) was added to each well except the wells containing either the protein standard or the resorufin standard. The plates were incubated for 10 min. NADPH (50 µl, 0.09 mM, Sigma-Aldrich) was added to all wells and the plates were incubated 10 min at room temperature. The deethylation reaction was stopped by adding 100 µl of 0.54 mM fluorescamine (in acetonitrile) to each well. The production of resorufin was measured in a fluorescence plate reader (TECANinfiniteM200, Tecan Austria GmbH, Grödig, Austria; excitation 544 nm, emission 590 nm) after 15 min. The EROD activity was expressed as picomole resorufin produced per milligramme protein per minute (pmol/(mg protein/min)). Protein was determined fluorometrically using the fluorescamine method (excitation 355 nm, emission 465 nm) [57,63]. Concentration-response curves and EC_5, EC_{10} and EC_{25} values were calculated using non-linear regression analyses of GraphPad Prism 5.0 (GraphPad, San Diego, USA).

AUTHORS' CONTRIBUTIONS

LG designed and conceived of the study, collected and prepared the samples, carried out the microtox, participated in the EROD assay and the Umu-C assay and drafted the manuscript. SH carried out the EROD assay. JE carried out the digestion part, and VR performed the Umu-C assay. HH participated in its design and coordination and helped to draft the manuscript. SK participated in the design of the study, supervised the study in Aachen and performed the statistical analysis. All authors read and approved the final manuscript.

ACKNOWLEDGEMENTS

We would like to thank Karlskoga Biogas for the financial support to this study. The authors would like to express their thanks to Drs. Niels C. Bols and Lucy Lee (University of Waterloo, Canada) for providing RTL-W1 cells.

REFERENCES

1.   Corselli-Nordblad Louise (2014) Renewable energy in the EU28. Eurostat Press Office 37: http://epp.eurostat.ec.europa.eu/cache/ITY_PUBLIC/8-10032014-AP/EN/8-10032014-AP-EN.PDF

2.   Energimyndigheten (Swedish authority for energy use) Produktion och användning av biogas år 2010 (Production and use of biogas during 2010). Report ES 2011:07.http://www.biogasvast.se/upload/Milj%C3%B6/Biogasproduktion%20Energimyndigheten%202011.pdf

3.   EU Council Directive 2009/28/ EC on the Promotion of the Use of Energy from Renewable Sources of the European Parliament and of the Council. http://europa.eu/legislation_summaries/energy/renewable_energy/en0009_en.htm

4.   EU Council (2011) COMMISSION STAFF WORKING DOCUMENT Recent Progress in Developing Renewable Energy Sources and Technical Evaluation of the use of Biofuels and Other Renewable Fuels in Transport in Accordance with Article

3 of Directive 2001/77/EC and Article 4(2) of Directive 2003/30/EC. http://europa.eu/legislation_summaries/energy/renewable_energy/l27065_en.htm

5. Jarvis Å, Schnürer A (2009) Mikrobiologisk handbok för biogasanläggningar. Swed Gas Technol Cent, Rapp SGC 207:1102-7371

6. Engwall M, Hjelm K (2000) Uptake of dioxin-like compounds from sewage sludge into various plant species–assessment of levels using a sensitive bioassay. Chemosphere 40:1189-1195

7. Hernández T, Moreno JI, Costa F (1991) Influence of sewage sludge application on crop yield and heavy metal availability. Soil Sci Plant Nutr 37:201-10

8. Hooda PS, McNulty D, Alloway BJ, Aitken MN (1997) Plant availability of heavy metals in soils previously amended with heavy applications of sewage sludge. J Sci Food Agricul 73:446-54

9. Corey RB, King LD, Lue-Hing C, Flanning DS (1987) Effects of sludge properties on accumulation of trace elements by crops. In: Page AL, Logan TJ, Ryan JA (eds) Land Application of Sludge – Food Chain Implications, Chelsea MI (US): Lewis Publisher.

10. Angelidaki I, Ahring BK (1997) Codigestion of olive oil mill wastewaters with manure, household waste or sewage sludge. Biodegradation 8:221-226

11. Chan YSG, Chu LM, Wong MH (1999) Codisposal of municipal refuse, sewage sludge and marine dredgings for methane production. Environ Pollut 106:123-128

12. Marques IP (2001) Anaerobic digestion treatment of olive mill wastewater for effluent re-use in irrigation. Desalination 137:233-239

13. Murto M, Björnsson L, Mattiasson B (2004) Impact of food industrial waste on anaerobic co-digestion of sewage sludge and pig manure. J Environ Manage 70:101-107

14. Sampaio MA, Gonçalves MR, Marques IP (2011) Anaerobic digestion challenge of raw olive mill wastewater. Bioresour technol 102:10810-10818

15. Heger S, Bluhm K, Agler MT, Maletz S, Schaeffer A, Seiler T-B, Angenent LT, Hollert H (2012) Biotests for hazard assessment of biofuel fermentation. Energ & Environ Sci 5:9778-9788

16. Kadlec RH, Knight RL (1996) Treatment Wetlands. London, New York: CRC Press.

17. Nielsen S, Willoughby N (2005) Sludge treatment and recycling of sludge and environmental impact. Topic in 10th European Biosolids and Biowaste Conference UK, November 2005. Water Environ J 19:285-296

18. Ugetti E, Ferrer I, Molist J, García J (2011) Technical, economic and environmental assessment of sludge treatment wetlands. Water Res 45:573-582

19. Bialowiec A, Randerson PF (2010) Phytotoxicity of landfill leachate on willow - Salix amygdalina L. Waste Manage 30:1587-1593

20. Kengne IM, Dodane P-H, Akoa A, Kone D (2010) Vertical-flow constructed wetlands as sustainable sanitation approach for faecal sludge dewatering in developing countries. Desalination 251:291-297

21. Merlin G, Pajean J-L, Lissolo T (2002) Performance of constructed wetlands for municipal wastewater treatment in rural mountainous area. Hydrobiologia 469:87-98

22. Nielsen S (2003) Sludge drying reedbeds. Wat Sci and Tec 48:101-109

23. Troesch S, Liénard A, Molle P, Merlin G, Esser D (2009) Sludge drying reed beds: s full and pilot-scales study for activated sludge treatment. Wat Sci Tech 60:1145-1154

24. Ugetti E, Ferrer I, Llorens E, García J (2010) Sludge treatment wetlands: a review on the state of the art. Bioresour Tech 101:2905-2912

25. Coleman J, Hench K, Garbutt K, Sexstone A, Bissonette G, Skousen J (2001) Treatment of domestic wastewater by three plant species in constructed wetlands. Water Air Soil Poll 128:283-285

26. Decamp O, Warren A (1998) Bacteriovory in ciliates isolated from constructed wetlands (reed beds) used for wastewater treatment. Water Res 12:1989-1996

27. Decamp O, Warren A (2000) Investigation of Escherichia coli removal in various designs of subsurface flow wetlands used for wastewater treatment. Ecol Eng 14:293-299

28. Moshiri GA (1993) Constructed wetlands for water quality

improvement. Michigan, USA: Lewis Publishers.

29. Gustavsson L, Engwall M (2012) Treatment of sludge containing nitro-aromatic compounds in reed-bed mesocosms - water, BOD, carbon and nutrient removal. Waste Manage 32:104-109

30. Gustavsson L, Hollert H, Jönsson S, van Bavel B, Engwall M (2007) Reed beds receiving industrial sludge containing nitroaromatic compounds - effects of outgoing water and bed material extracts in the umu-C genotoxicity assay, DR-CALUX assay and on early life stage development in zebrafish (danio rerio). Environ Sci Pollut Res 14:202-211

31. Gustavsson L, Engwall M, Jönsson S, Van Bavel B (2003) Biological treatment of sludge containing explosives and pharmaceutical residues – effects on toxicity and chemical concentrations. In: Third Conference of Disposal of Energetic Material. Karlskoga: KCEM (Knowledge Centre for Energetic Materials) - Section of Detonation and Combustion.

32. Klee N, Gustavsson L, Kosmehl T, Engwall M, Erdinger L, Braunbeck T, Hollert H (2004) Changes in toxicity and genotoxicity of industrial sewage sludge samples containing nitro- and amino-aromatic compounds following treatment in bioreactors with different oxygen regimes. ESPR, Environ Sci Pollut Res 11:313-320

33. Gustavsson L, Klee N, Olsman H, Hollert H, Engwall M (2004) Fate of Ah receptor agonists during biological treatment of an industrial sludge containing explosives and pharmaceutical residues. ESPR, Environ Sci Pollut Res 11:379-386

34. Gustavsson L, Engwall M (2006) Genotoxic activity of nitroarene-contaminated industrial sludge following large-scale treatment in aerated and non-aerated sacs. Sci Tot Env 367:694-703

35. Swedish EPA (2009) Riktvärden för förorenad mark. NV Rapport. 5976

36. Swedish Waste Association (2008) Den Svenska Biogaspotentialen Från Inhemska Varor. 2: Report 2008

37. Swedish Waste Association (Avfall Sverige Utveckling) (2009) Substrathandbok för Biogasproduktion. 14: Report U2009

38. Ku çu ÖS, Sponza DT (2009) Kinetics of para-nitrophenol and chemical oxygen demand removal from synthetic wastewater in

an anaerobic migrating blanket reactor. J Hazard Mater 161:787-799

39. Hecht C, Griehl C (2009) Investigation of the accumulation of aromatic compounds during biogas production from kitchen waste. Bioresour Tech 100:654-658

40. Ku çu ÖS, Sponza DT (2009) Effect of increasing nitrobenzene loading rates on the performance of anaerobic migrating blanket reactor and sequential anaerobic migrating blanket reactor/completely stirred tank reactor system. J Hazard Mater 168:390-399

41. Karathanasis AD, Potter CL, Coyone MS (2003) Vegetation effects on fecal bacteria, BOD and suspended solid removal in constructed wetlands treating domestic wastewater. Ecol Eng 20:157-169

42. Luederitz V, Eckert E, Lange-Weber M, Lange A, Gersberg RM (2001) Nutrient removal efficiency and resource economics of vertical flow and horizontal flow constructed wetlands. Ecol Eng 18:157-171

43. Boopathy R, Kulpa C (1997) Anaerobic biodegradation of explosives and related compounds by sulfate-reducing and methanogenic bacteria; a review. Bioresour Tech 63:81-89

44. Padda RS, Wang CY (2000) Mutagenicity of trinitrotoluene and metabolites formed during anaerobic degradation by Clostridium acetobutylicum ATCC 824. Environ Toxicol Chem 19:2871-2875

45. Rogers JD, Bunce NJ (2001) Treatment methods for the remediation of nitroaromatic explosives. Water Res 35:2101-2111

46. Sundvall A, Marklund H, Rannug U (1984) The mutagenicity on Salmonella typhimurium of nitrobenzoic acids and other wastewater components generated in the production of nitrobenzoic acids and nitrotoluenes. Mutat Res 137:71-78 |

47. Boopathy R, Manning JF, Kupla CF (1998) A laboratory study of the bioremediation of 2,4,6-trinitrotoluene-contaminated soil using aerobic/anoxic soil slurry reactor. Water Environ Res 70:80-86

48. Razo-Flores E, Donlon B, Lettinga G, Field JA (1997) Biotransformation and biodegradation of N-substituted aromatics in methanogenic granular sludge. FEMS Microbiol Rev 20:525-538

49. Toze S, Zappia L (1999) Microbial degradation of munition compounds in production wastewater. Water Res 33:3040-3045

50. Umbuzeiro GA, Franco A, Martins MH, Kummrow F, Carvalho L, Schmeiser HH, Leykauf J, Stiborova M, Claxton LD (2008) Mutagenicity and DNA adduct formation of PAH, nitro-PAH, and oxy-PAH fractions of atmospheric particulate matter from São Paulo, Brazil. Mutat Res 652:172-80

51. Engwall M, Schnürer A (2002) Fate of Ah-receptor agonists in organic household waste during anaerobic degradation-estimation of levels using EROD induction in organic cultures of chick embryo livers. Sci Tot Environ 297:105-108

52. Olsman H, Björnfoth H, van Bavel B, Lindström G, Schnürer A, Engwall M (2002) Characterisation of dioxin-like compounds in anaerobically digested organic material by bioassay-directed fractionation. Organohalogen Compd 58:345-348

53. Chen I, Safe S, Bjeldanes L (1996) Indole-3-carbinol and diindolmethane as aryl hydrocarbon (Ah) receptor agonist and antagonist in T47D human breast cancer cells. Biochem Pharmacol 51:1069-1076

54. Naur P, Hansen HC, Bak S, Hansen GB, Jensen NB, Nielsen HL, Halkier BA (2003) CYP79B1 from *Sinapsis alba* coverts tryptophan to indole-3-acetaldoxime. Arch Biochem Biophys 409:235-241

55. Teel RW, Huynh H (1998) Modulation by phytochemicals of cytochrome P450 linked enzyme activity. Cancer Lett 133:135-141

56. Jonsson S, Borén H (2002) Analysis of mono- and diesters of o-phthalic acid by solid phase extractions with polystyrene-divinylbenzene-based polymers. J Chromatogr A 963:393-400

57. Hollert H, Dürr M, Olsman H, Halldin K, Bavel B v, Brack W, Tysklind M, Engwall M, Braunbeck T (2002) Biological and chemical determination of dioxin-like compounds in sediments by means of a sediment triad approach in the catchment area of the Neckar River. Ecotoxicology 11:323-336

58. Keiter S, Grund S, van Bavel B, Hagberg J, Engwall M, Kammann U, Klempt M, Manz W, Olsman H, Braunbeck T, Hollert H (2008) Activities and identification of aryl hydrocarbon receptor agonists in sediments from the Danube river. Anal Bioanal Chem 390:2009-2019

59. Olsman H, Hagberg J, Kalbin G, Julander A, van Bavel B, Strid Å, Tysklind M, Engwall M (2005) Ah receptor agonists in UV-exposed toluene solutions of decabromodiphenyl ether (decaBDE) and in soils contaminated with polybrominated diphenyl ethers (PBDEs). ESPR - Environ Sci Pollut Res 13:161-169

60. Lee LE, Clemons JH, Bechtel DG, Caldwell SJ, Han KB, Pasitschniak-Arts M, Mosser D, Bols NC (1993) Development and characterization of a rainbow trout liver cell line expressing cytochrome p450-dependent monooxygenase activity. Cell Bio Toxicol 9:279-294

61. Behrens A, Schirmer K, Bols NC, Segner H (2001) Polycyclic aromatic hydrocarbons as inducers of cytochrome P4501A enzyme activity in the rainbow trout liver cell line, RTL-W1, and in primary cultures of rainbow trout hepatocytes. Environ Toxicol Chem 20:632-643

62. Seiler TB, Rastall AC, Leist E, Erdinger L, Braunbeck T, Hollert H (2006) Membrane dialysis extraction (MDE): a novel approach for extracting toxicologically relevant hydrophobic organic compounds from soils and sediments for assessment in biotests. J Soils Sediments 6:20-29

63. Brunström B, Halldin K (1998) EROD induction by environmental contaminants in avian embryo livers. Comp Biochem Physiol C Pharmacol Toxicol Endocrinol 121:213-219

Citations

CHAPTER 1

Jenő Hancsók, Péter Baladincz, Tamás Kasza, Sándor Kovács, Csaba Tóth, and Zoltán Varga, "Bio Gas Oil Production from Waste Lard,"Journal of Biomedicine and Biotechnology, vol. 2011, Article ID 384184, 9 pages, 2011. doi:10.1155/2011/384184.

CHAPTER 2

Chukwuka G. Monyei, Aderemi O. Adewumi, and Michael O. Obolo, "Oil Well Characterization and Artificial Gas Lift Optimization Using Neural Networks Combined with Genetic Algorithm," Discrete Dynamics in Nature and Society, vol. 2014, Article ID 289239, 10 pages, 2014. doi:10.1155/2014/289239.

CHAPTER 3

Birte Viétor, Thomas Hoppe, and Joy Clancy, Decentralised Combined Heat and Power in the German Ruhr Valley; Assessment of Factors Blocking Uptake and Integration, doi:10.1186/s13705-015-0033-0.

CHAPTER 4

Tomohiro Toki, Ryosaku Higa, Akira Ijiri, Urumu Tsunogai, and Juichiro Ashi, Origin and Transport of Pore Fluids in the Nankai Accretionary Prism Inferred from Chemical and Isotopic Compositions of Pore Water at Cold Seep Sites off Kumano,doi:10.1186/s40623-014-0137-3.

CHAPTER 5

Xin Li, Zhi-Gang Li, and Zhong-Ping Shi, Metabolic flux and transcriptional analysis elucidate higher butanol/acetone ratio feature in ABE extractive fermentation by Clostridium acetobutylicum using cassava substrate, doi:10.1186/s40643-014-0013-9.

CHAPTER 6

Emerson Léo Schultz, Daniela Tatiane de Souza,and Mônica Caramez Triches Damaso, The glycerol biorefinery: a purpose for Brazilian biodiesel production, doi:10.1186/s40538-014-0007-z.

CHAPTER 7

Hodaka Kawahata, Shusuke Yamashita, Kyoko Yamaoka, Takashi Okai, Gen Shimoda, and Noboru Imai, Heavy Metal Pollution in Ancient Nara, Japan, During the Eighth Century, doi:10.1186/2197-4284-1-15.

CHAPTER 8

Lillemor K Gustavsson, Sebastian Heger, Jörgen Ejlertsson, Veronica Ribé, Henner Hollert, and Steffen H Keiter, Industrial sludge containing pharmaceutical residues and explosives alters inherent toxic properties when co-digested with oat and post-treated in reed beds, doi:10.1186/2190-4715-26-8.

Index